PROOFS AND CONFIRMATIONS

This is an introduction to recent developments in algebraic combinatorics and an illustration of how research in mathematics actually progresses. The author recounts the story of the search for and discovery of a proof of a formula conjectured in the early 1980s: the number of $n \times n$ alternating sign matrices, objects that generalize permutation matrices. Although it was soon apparent that the conjecture must be true, the proof was elusive. Researchers became drawn to this problem, making connections to aspects of the invariant theory of Jacobi, Sylvester, Cayley, MacMahon, Schur, and Young, to partitions and plane partitions, to symmetric functions, to hypergeometric and basic hypergeometric series, and, finally, to the six-vertex model of statistical mechanics. All these threads are brought together in Zeilberger's 1995 proof of the original conjecture.

The book is accessible to anyone with a knowledge of linear algebra. Students will learn what mathematicians actually do, and even researchers in combinatorics will find something new here.

David M. Bressoud is DeWitt Wallace Professor and Chair of Mathematics and Computer Science at Macalester College, St. Paul, Minnesota. He was a Peace Corps Volunteer in Antigua, West Indies, received his Ph.D. from Temple University, and taught at The Pennsylvania State University before moving to Macalester. He has held visiting positions at the Institute for Advanced Study, University of Wisconsin, University of Minnesota, and the University of Strasbourg. He has received a Sloan Fellowship, a Fulbright Fellowship, and the MAA Distinguished Teaching Award. He has published more than fifty research articles in number theory, partition theory, combinatorics, and the theory of special functions. His other books include *Factorization and Primality Testing*, *Second Year Calculus from Celestial Mechanics to Special Relativity*, and *A Radical Approach to Real Analysis*.

SPECTRUM SERIES

Published by
THE MATHEMATICAL ASSOCIATION OF AMERICA

SPECTRUM SERIES

The Spectrum Series of the Mathematical Association of America was so named to reflect its purpose: to publish a broad range of books including biographies, accessible expositions of old or new mathematical ideas, reprints and revisions of excellent out-of-print books, popular works, and other monographs of high interest that will appeal to a broad range of readers, including students and teachers of mathematics, mathematical amateurs, and researchers.

All the Math That's Fit to Print, by Keith Devlin
Circles: A Mathematical View, by Dan Pedoe
Complex Numbers and Geometry, by Liang-shin Hahn
Cryptology, by Albrecht Beutelspacher
Five Hundred Mathematical Challenges, by Edward J. Barbeau, Murray S. Klamkin, and William O. J. Moser
From Zero to Infinity, by Constance Reid
I Want To Be a Mathematician, by Paul R. Halmos
Journey into Geometries, by Marta Sved
JULIA: A Life in Mathematics, by Constance Reid
The Last Problem, by E. T. Bell (revised and updated by Underwood Dudley)
The Lighter Side of Mathematics: Proceedings of the Eugène Strens Memorial Conference on Recreational Mathematics & Its History, edited by Richard K. Guy and Robert E. Woodrow
Lure of the Integers, by Joe Roberts
Magic Tricks, Card Shuffling, and Dynamic Computer Memories: The Mathematics of the Perfect Shuffle, by S. Brent Morris
Mathematical Carnival, by Martin Gardner
Mathematical Circus, by Martin Gardner
Mathematical Cranks, by Underwood Dudley
Mathematical Magic Show, by Martin Gardner
Mathematics: Queen and Servant of Science, by E. T. Bell
Memorabilia Mathematica, by Robert Edouard Moritz
New Mathematical Diversions, by Martin Gardner
Non-Euclidean Geometry, by H. S. M. Coxeter
Numerical Methods That Work, by Forman Acton
Numerology or What Pythagoras Wrought, by Underwood Dudley
Out of the Mouths of Mathematicians, by Rosemary Schmalz
Penrose Tiles to Trapdoor Ciphers . . . and the Return of Dr. Matrix, by Martin Gardner
Polyominoes, by George Martin
The Search for E. T. Bell, Also Known as John Taine, by Constance Reid
Shaping Space, edited by Marjorie Senechal and George Fleck
Student Research Projects in Calculus, by Marcus Cohen, Edward D. Gaughan, Arthur Knoebel, Douglas S. Kurtz, and David Pengelley
The Trisectors, by Underwood Dudley
Twenty Years Before the Blackboard, by Michael Stueben with Diane Sandford
The Words of Mathematics, by Steven Schwartzman

PROOFS AND CONFIRMATIONS

The Story of the Alternating Sign Matrix Conjecture

DAVID M. BRESSOUD
Macalester College
St. Paul, Minnesota

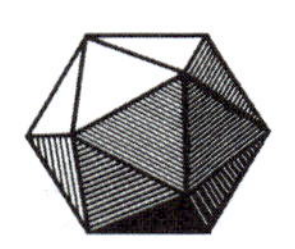

THE MATHEMATICAL ASSOCIATION OF AMERICA

COPUBLISHED BY THE PRESS SYNDICATE OF THE UNIVERSITY OF CAMBRIDGE
The Pitt Building, Trumpington Street, Cambridge, United Kingdom
And the Mathematical Association of America

CAMBRIDGE UNIVERSITY PRESS
The Edinburgh Building, Cambridge CB2 2RU, UK www.cup.cam.ac.uk
40 West 20th Street, New York, NY 10011-4211, USA www.cup.org
10 Stamford Road, Oakleigh, Melbourne 3166, Australia
Ruiz de Alarcón 13, 28014 Madrid, Spain
MATHEMATICAL ASSOCIATION OF AMERICA
1529 Eighteenth St. NW, Washington DC 20036 www.maa.org

First published 1999

Printed in the United States of America

Typeset by the author

A catalog record for this book is available from the British Library.

Library of Congress Cataloging-in-Publication Data is available.

ISBN 0 521 66170 6 hardback
ISBN 0 521 66646 5 paperback

this book is dedicated to

Janet Alford

friend and mother-in-law
and to

George Andrews *and* Dick Askey

mentors who have educated my taste in mathematics

Contents

Preface

Mathematics does not grow through a monotonous increase of the number of indubitably established theorems but through the incessant improvement of guesses by speculation and criticism, by the logic of proofs and refutations.

– Imre Lakatos (1976)

What is the role of proof in mathematics? In recent years we have seen a rebellion against teaching mathematics as a procession of irrefutable logical arguments that build from fundamental principles to universal truths. In this age of computer exploration of the patterns of mathematics, the pendulum has swung so far that some have proclaimed "The Death of Proof" (Horgan 1993). Traditional proofs are disappearing from the high school and even early college curricula.

This loss should not be mourned if what we are giving up is the misconception of mathematics as a formal system without need for scientific exploration, experimentation, and discovery. But there is a very real danger that in our enthusiasm to jettison what is false, we will lose the very essence of mathematics which is inextricably tied to proofs. We are brought back to our initial question: What *is* the role of proof in mathematics?

Imre Lakatos in *Proofs and Refutations* (1976) has highlighted one important aspect of this role. His is an ironic insight, for what is proven should be true and therefore not refutable. Lakatos's inspiration was Karl Popper, and the title is adapted from Popper's *Conjectures and Refutations* (1963). Popper's thesis was that science does not progress by establishing truth, but rather by advancing conjectures that can be tested against reality, that are maintained so long as they agree with reality, and that are refined or rejected when they fail in their predictions.

Lakatos tackled the question of whether this understanding also applies to mathematics. On the surface, Popper's work seems irrelevant. There is nothing provisional about the Pythagorean theorem. There is no possibility that someone will someday discover a right triangle for which the sums of the squares of the legs is not equal to the square of the hypotenuse. Or is there?

In fact, there are metric spaces with their associated geometries in which the Pythagorean theorem is not true. This is not a fair attack because the Pythagorean theorem is intended to apply only to Euclidean geometry with the standard Euclidean metric, but this is precisely Lakatos's point: An important part of the dynamic of mathematics is the ferreting out of these underlying assumptions and the subsequent refinement of the mathematical concepts with which we work.

Lakatos illustrated this process by recounting the history of Euler's formula relating the number of vertices, edges, and faces of a polyhedron. In an appendix, he mentions the relevance of this viewpoint to the development of the concept of uniform continuity. I drew on his insight when I wrote *A Radical Approach to Real Analysis*. Mathematics comes alive as we see how seemingly esoteric definitions arise from the struggle to reconcile the perceived patterns of mathematics with the counterexamples that are encountered in the recognition and delimitation of our assumptions.

But this is not the whole story of the role of proof in mathematics. There is another dynamic at work that has little to do with the precision of the definitions or the boundaries of the assumptions. It involves the search for proof which is at the core of what a mathematician does. In his Rouse Ball lecture of 1928, G. H. Hardy expressed his view of the nature of mathematical proof. The italics are his.

> I have myself always thought of a mathematician as in the first instance an *observer*, a man who gazes at a distant range of mountains and notes down his observations. His object is simply to distinguish clearly and notify to others as many different peaks as he can. There are some peaks which he can distinguish easily, while others are less clear. He sees A sharply, while of B he can obtain only transitory glimpses. At last he makes out a ridge which leads from A, and following it to its end he discovers that it culminates in B. B is now fixed in his vision, and from this point he can proceed to further discoveries. In other cases perhaps he can distinguish a ridge which vanishes in the distance, and conjectures that it leads to a peak in the clouds or below the horizon. But when he sees a peak he believes that it is there simply because he sees it. If he wishes someone else to see it, he *points to it*, either directly or through the chain of summits which led him to recognize it himself. When his pupil also sees it, the research, the argument, the *proof* is finished.

The analogy is a rough one, but I am sure that it is not altogether misleading. If we were to push it to its extreme we should be led to a rather paradoxical conclusion; that we can, in the last analysis, do nothing but *point*; that proofs are what Littlewood and I call *gas*, rhetorical flourishes designed to affect psychology, pictures on the board in the lecture, devices to stimulate the imagination of pupils. This is plainly not the whole truth, but there is a good deal in it. The image gives us a genuine approximation to the processes of mathematical pedagogy on the one hand and of mathematical discovery on the other; it is only the very unsophisticated outsider who imagines that mathematicians make discoveries by turning the handle of some miraculous machine. Finally the image gives us at any rate a crude picture of Hilbert's metamathematical proof, the sort of proof which is a *ground* for its conclusion and whose object is to *convince*.

I want to push Hardy's metaphor in a different direction, to exhibit proofs as a form of exploration. There are circumstances in which it is possible to stand far off and indicate the succession of ridges that will lead to the peak and so establish, definitively, the route to that highest point. More often, the path to the summit can be discovered only by going out and breaking it. This is a process that involves false starts, dense fog, and a lot of hard work. It is aided by an intuitive feel for the lay of the land, but it is almost always filled with surprises.

Mathematicians often recognize truth without knowing how to prove it. Confirmations come in many forms. Proof is only one of them. But knowing something is true is far from understanding why it is true and how it connects to the rest of what we know. The search for proof is the first step in the search for understanding.

I can best illustrate my point by describing real mathematics. I shall do so with a story of recent discoveries at the boundary of algebra and combinatorics, the story of the alternating sign matrix conjecture proposed by William Mills, David Robbins, and Howard Rumsey. For fifteen years this conjecture was known to be true. It was validated in many ways: in the surprising simplicity of its formulation, by numerical verification of the first twenty cases, through unanticipated implications that could be proven. But none of this was proof. When a proof finally was found in 1995, it lay in unexpected territory and revealed a host of new insights and engaging problems.

My intention in this book is not just to describe this discovery of new mathematics, but to guide you into this land and to lead you up some of the recently scaled peaks. This is not an exhaustive account of all the marvels that have been uncovered, but rather a selected tour that will, I hope, encourage you to return and pursue your own explorations.

The narrative of my story is contained in Chapters 1 and 6, with concluding comments in Section 7.4. It is possible to read this and nothing more, but even casual readers would benefit from an inspection of Chapters 2 and 3. These chapters provide the context for this story and examine the fundamental results on which the study of the alternating sign matrix conjecture builds: generating functions, recursive formulæ, partitions, plane partitions, lattice paths, and determinant evaluations. At the base of this story lies the nineteenth-century invariant theory of Cauchy, Jacobi, Cayley, Sylvester, and MacMahon. Even down these well-traveled paths, we shall find results and interpretations that were discovered only recently. This foundational material concludes in Section 3.5 with a description and explanation of Dodgson's algorithm for evaluating determinants, the mathematics that was to lead Robbins and Rumsey to their discovery of alternating sign matrices.

Chapters 4, 5, and 7 are each devoted to a single proof. In Chapter 4, we follow Ian Macdonald's route through the theory of symmetric functions to the proof of a result first conjectured by MacMahon, the generating function for bounded, symmetric plane partitions. It was from this vantage that Macdonald spied our second peak, his conjectured generating function for bounded, cyclically symmetric plane partitions.

Chapter 5 includes the proof of Macdonald's conjecture, a true example of mathematical serendipity. As Mills, Robbins, and Rumsey sought to prove their conjecture, they stumbled upon a connection to plane partitions and Macdonald's conjecture. They had come upon it from a new direction and so came armed with an additional piece of information that proved to be crucial. This chapter also provides an opportunity to explore the world of basic hypergeometric series, another ingredient in the ultimate proof of the alternating sign matrix conjecture.

The proof of Mills, Robbins, and Rumsey's original conjecture was found by Doron Zeilberger in 1995. His proof is presented in Chapter 7. It traverses ground that was first opened up by physicists working in statistical mechanics and draws on the theory of orthogonal polynomials.

The three main proofs are difficult, especially for anyone who has never before tackled a complicated proof, but the only requirement to set out is a familiarity with linear algebra and a desire to glimpse a piece of current mathematical research. The exercises are designed to help you prepare, as well as to point out some of the features and views along the way. I have scouted out what I feel are the easiest ascents, but none of them is easy, not if you are really going to make them. This means not just agreeing that the route is legitimate but coming to your own understanding of the ideas that are involved and the reasons why these theorems are correct. This is what mathematical proof is really about. It is not just the confirmation of the validity of a mathematical insight. Very early in their quest, Mills, Robbins, and Rumsey knew that their alternating sign matrix conjecture was true. They sought a proof because they wanted to understand why it was true and to see where that understanding might take them.

A word on terminology. A **conjecture** is a statement that is believed but not proven to be true. Strictly speaking, the alternating sign matrix conjecture no longer exists. It is now the alternating sign matrix theorem. Since part of the purpose of this story is to explain the role of certain conjectures in recent mathematical developments, I have decided to speak of these results in terms of their historical status as conjectures. In making this decision, I take on the responsibility to clearly identify which conjectures have been proven.

I am indebted to many people who have read early versions of this book and made helpful suggestions for its improvement. Chief among these are Art Benjamin and Jennifer Quinn, who went through the entire manuscript and caught many of my errors. I especially wish to thank Krishna Alladi, George Andrews, Dick Askey, Rodney Baxter, Dominique Foata, Tina Garrett, Ira Gessel, Gouniou Han, Christian Krattenthaler, Greg Kuperberg, Alain Lascoux, Ina Lindemann, Ian Macdonald, Peter Paule, Jim Propp, David Robbins, Richard Stanley, Dennis Stanton, Xavier Viennot, and Doron Zeilberger. I also wish to thank Keith Dennis, Julio Gonzalez Cabillon, Karen Parshall, Bob Proctor, and Jim Tattersall for references. Finally, I owe a debt of gratitude to Lauren Cowles at Cambridge University Press and Don Albers at the Mathematical Association of America for their continued support and encouragement.

The jigsaw puzzle cover art was created by Greg Kuperberg and is reproduced with his permission. The picture of Cauchy is published by permission of the American Institute of Physics Emilio Segrè Visual Archives. The picture of Sylvester is published by permission of the Ferdinand Hamburger, Jr. Archives of the Johns Hopkins University. The picture of Jacobi is courtesy of the Matematisches Forschungsinstitut Oberwolfach. The picture of MacMahon is courtesy of George Andrews.

David M. Bressoud, www.macalester.edu/~bressoud
Macalester College, St. Paul, Minnesota

1
The Conjecture

It is difficult to give an idea of the vast extent of modern mathematics. This word "extent" is not the right one: I mean extent crowded with beautiful detail – not an extent of mere uniformity such as an objectless plain, but of a tract of beautiful country seen at first in the distance, but which will bear to be rambled through and studied in every detail of hillside and valley, stream, rock, wood, and flower. But, as for everything else, so for a mathematical theory – beauty can be perceived but not explained.

– Arthur Cayley (1883)

Conjectures are the warp upon which we weave mathematics. They are not to be confused with guesses. André Weil is reported to have proclaimed in exasperation at a guess that its author had dared elevate to the rank of conjecture, "*That* is not a conjecture; *that* is just talking."

A conjecture may be false, but this must come as a surprise, as an illumination revealing unimagined subtleties. A good conjecture comes with the certainty that it must be true: because it simplifies what had seemed complicated and brings order to what had appeared chaotic, because it carries implications which themselves seem right and may even be verifiable, because it bears the stamp of elegance that the trained observer has learned to recognize as the hallmark of truth.

This story is woven on the strands of fourteen conjectures that I have chosen from the many that arose in the course of investigations of alternating sign matrices. All but two of these chosen fourteen have been proven. Many other related conjectures are still open problems. We shall see the details of many of the proofs, but they are less important than the themes that they draw upon. This is the power of a good conjecture and the reason for seeking its proof: It can reveal unexpected connections and guide us to the ideas that are truly important.

In the early 1980s, William Mills, David Robbins, and Howard Rum-

Figure 1.1. (*Left to right*) William Mills, David Robbins, and Howard Rumsey.

sey made a good conjecture. It arose from Robbins and Rumsey's study of an algorithm for evaluating determinants, an algorithm that had been discovered more than a century earlier by the Reverend Charles Lutwidge Dodgson, better known as Lewis Carroll. Dodgson's technique proceeds by iteratively evaluating 2×2 determinants, and it suggested a certain generalization of the determinant. While an ordinary determinant of an $n \times n$ matrix can be expressed as a sum over the permutations of n letters, Robbins and Rumsey's extraordinary determinant is a sum over alternating sign matrices. The story of the genesis of alternating sign matrices will be told in Section 3.5. Here we are going to focus on the simple question, how many are there?

A **permutation** of n letters is a one-to-one mapping, which we shall usually denote by σ, from the set $\{1, 2, \ldots, n\}$ into itself. An example of a permutation of 5 letters is given by

$$\sigma(1) = 1, \quad \sigma(2) = 3, \quad \sigma(3) = 5, \quad \sigma(4) = 2, \quad \sigma(5) = 4.$$

We shall denote the set of permutations of n letters by $\mathcal{S}_n$. It is a group for which composition acts as multiplication

$$\sigma \circ \tau(j) = \sigma(\tau(j)). \tag{1.1}$$

A permutation on n letters can be represented by an $n \times n$ matrix of

0s and 1s. For row i, we put a 1 in column $\sigma(i)$ and a zero in every other column. For our example, the corresponding matrix is

$$\begin{pmatrix} 1 & 0 & 0 & 0 & 0 \\ 0 & 0 & 1 & 0 & 0 \\ 0 & 0 & 0 & 0 & 1 \\ 0 & 1 & 0 & 0 & 0 \\ 0 & 0 & 0 & 1 & 0 \end{pmatrix}.$$

Any square matrix of 0s and 1s with exactly one 1 in each row and in each column represents a permutation. This representation is particularly useful because composition corresponds to matrix multiplication. In other words, in order to multiply two elements of $\mathcal{S}_n$, we multiply the corresponding matrices, although as we have defined the correspondence, it is necessary to reverse the order of multiplication (see exercise 1.1.3).

An **alternating sign matrix** is a square matrix of 0s, 1s, and -1s for which

- the sum of the entries in each row and in each column is 1,
- the non-zero entries of each row and of each column alternate in sign.

An example of such a matrix is

$$\begin{pmatrix} 0 & 1 & 0 & 0 & 0 \\ 0 & 0 & 1 & 0 & 0 \\ 1 & -1 & 0 & 0 & 1 \\ 0 & 1 & -1 & 1 & 0 \\ 0 & 0 & 1 & 0 & 0 \end{pmatrix}.$$

Any permutation matrix is also an alternating sign matrix. We know that the number of $n \times n$ permutation matrices is $n!$. Robbins and Rumsey wanted to know how much larger the number of $n \times n$ alternating sign matrices would be.

1.1 How many are there?

We let A_n denote the number of $n \times n$ alternating sign matrices. In Chapter 2 we shall see how we can count $n \times n$ alternating sign matrices efficiently. The numbers that we get suggest that A_n is much larger than $n!$:

n	$n!$	A_n
1	1	1
2	2	2
3	6	7
4	24	42
5	120	429
6	720	7436
7	5040	218348
8	40320	10850216
9	362880	911835460
$\vdots$	$\vdots$	$\vdots$

While $16! = 20922789888000 \approx 2.1 \times 10^{13}$, we have that

$$A_{16} = 64427185703425689356896743840 \approx 6.4 \times 10^{28}.$$

The difference is substantial. Exactly how much faster does A_n grow? To answer this, we would like to find a closed formula for A_n. David Robbins had the crucial idea that was to enable them to find such a formula:

When faced with combinatorial enumeration problems, I have a habit of trying to make the data look similar to Pascal's triangle. There is a natural way to do this with alternating sign matrices. One sees easily that every alternating sign matrix has a single 1 in its top row. (Robbins 1991)

We cannot have any -1s in the first row of an alternating sign matrix (why not?), which is why there is exactly one 1 in this row. We use this observation to split the $n \times n$ alternating sign matrices into n subsets according to the position of the 1 in the first row. For example, if $n = 3$, then there are two matrices with a 1 at the top of the first column:

$$\begin{pmatrix} 1 & 0 & 0 \\ 0 & 1 & 0 \\ 0 & 0 & 1 \end{pmatrix} \quad \text{and} \quad \begin{pmatrix} 1 & 0 & 0 \\ 0 & 0 & 1 \\ 0 & 1 & 0 \end{pmatrix},$$

three matrices with a 1 at the top of the second column:

$$\begin{pmatrix} 0 & 1 & 0 \\ 1 & 0 & 0 \\ 0 & 0 & 1 \end{pmatrix}, \quad \begin{pmatrix} 0 & 1 & 0 \\ 1 & -1 & 1 \\ 0 & 1 & 0 \end{pmatrix}, \quad \text{and} \quad \begin{pmatrix} 0 & 1 & 0 \\ 0 & 0 & 1 \\ 1 & 0 & 0 \end{pmatrix},$$

and two matrices with a 1 at the top of the third column:

$$\begin{pmatrix} 0 & 0 & 1 \\ 1 & 0 & 0 \\ 0 & 1 & 0 \end{pmatrix} \quad \text{and} \quad \begin{pmatrix} 0 & 0 & 1 \\ 0 & 1 & 0 \\ 1 & 0 & 0 \end{pmatrix}.$$

If we define $A_{n,k}$ to be the number of $n \times n$ alternating sign matrices with a 1 at the top of the kth column, then we have

$$A_{3,1} = 2, \qquad A_{3,2} = 3, \qquad A_{3,3} = 2.$$

We can now construct something like Pascal's triangle in which the kth entry of the nth row is $A_{n,k}$:

$$\begin{array}{ccccccccccc} & & & & & 1 & & & & & \\ & & & & 1 & & 1 & & & & \\ & & & 2 & & 3 & & 2 & & & \\ & & 7 & & 14 & & 14 & & 7 & & \\ & 42 & & 105 & & 135 & & 105 & & 42 & \\ 429 & & 1287 & & 2002 & & 2002 & & 1287 & & 429. \end{array}$$

There are two observations about this triangular array that are easy to verify. The first is that it is symmetric:

$$A_{n,k} = A_{n,n+1-k}.$$

This follows from the fact that the vertical mirror image of an alternating sign matrix is another alternating sign matrix. The second observation is that the numbers that run down the outer edge count the total number of alternating sign matrices of the previous size. In other words,

$$A_{n,1} = A_{n,n} = A_{n-1}.$$

If we have a 1 in the first column of the first row, then all other entries in the first row and the first column must be zero, and the number of ways of filling in the remainder of the matrix is the number of $n-1 \times n-1$ alternating sign matrices:

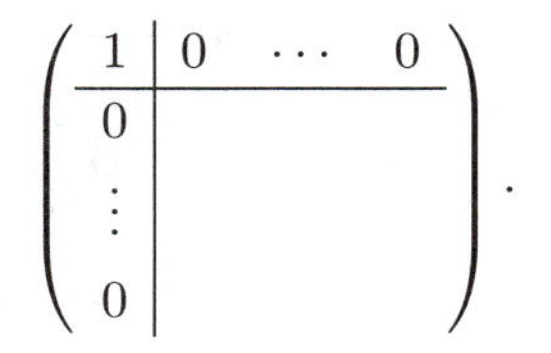

By now, Robbins and Rumsey had drawn William Mills into their investigations and had taken the next creative step which was to look at the ratios of consecutive entries in the triangle. After playing around with different possible representations for each ratio, the following pattern emerged:

$$
\begin{array}{ccccccccccccccccc}
 & & & & & & & & 1 & & & & & & & & \\
 & & & & & & 1 & & \mathbf{2/2} & & 1 & & & & & & \\
 & & & & 2 & & \mathbf{2/3} & & 3 & & \mathbf{3/2} & & 2 & & & & \\
 & & 7 & & \mathbf{2/4} & & 14 & & \mathbf{5/5} & & 14 & & \mathbf{4/2} & & 7 & & \\
 & 42 & \mathbf{2/5} & 105 & & \mathbf{7/9} & & 135 & & \mathbf{9/7} & & 105 & & \mathbf{5/2} & 42 & & \\
429 & \mathbf{2/6} & 1287 & \mathbf{9/14} & 2002 & & \mathbf{16/16} & & 2002 & & \mathbf{14/9} & 1287 & \mathbf{6/2} & 429. & & &
\end{array}
$$

The ratios themselves form a curious Pascal's triangle

$$
\begin{array}{ccccccccccc}
 & & & & & \mathbf{2/2} & & & & & \\
 & & & & \mathbf{2/3} & & \mathbf{3/2} & & & & \\
 & & & \mathbf{2/4} & & \mathbf{5/5} & & \mathbf{4/2} & & & \\
 & & \mathbf{2/5} & & \mathbf{7/9} & & \mathbf{9/7} & & \mathbf{5/2} & & \\
 & \mathbf{2/6} & & \mathbf{9/14} & & \mathbf{16/16} & & \mathbf{14/9} & & \mathbf{6/2} & \\
\mathbf{2/7} & & \mathbf{11/20} & & \mathbf{25/30} & & \mathbf{30/25} & & \mathbf{20/11} & & \mathbf{7/2}
\end{array}
$$

with $2/(n+1)$ as the first entry in the nth row, $(n+1)/2$ as the last entry, and all intermediate entries formed by adding numerators and adding denominators of the two entries directly above it. For example,

$$\frac{A_{5,1}}{A_{5,2}} = \frac{2}{5}, \qquad \frac{A_{5,2}}{A_{5,3}} = \frac{7}{9}$$

and the second ratio of line 6 is

$$\frac{A_{6,2}}{A_{6,3}} = \frac{9}{14} = \frac{2+7}{5+9}.$$

This strange addition of ratios suggests that the numerators and denominators are each a superposition of Pascal triangles. In fact, once we look for it we see that the numerators naturally decompose as

$$
\begin{array}{ccccccccccc}
 & & & & & 1+1 & & & & & \\
 & & & & 1+1 & & 1+2 & & & & \\
 & & & 1+1 & & 2+3 & & 1+3 & & & \\
 & & 1+1 & & 3+4 & & 3+6 & & 1+4 & & \\
 & 1+1 & & 4+5 & & 6+10 & & 4+10 & & 1+5 & \\
1+1 & & 5+6 & & 10+15 & & 10+20 & & 5+15 & & 1+6
\end{array}
$$

This leads to the following conjecture whose proof is the ultimate object of this book.

Conjecture 1 (The refined ASM conjecture) *For* $1 \le k < n$,

$$\frac{A_{n,k}}{A_{n,k+1}} = \frac{\binom{n-2}{k-1} + \binom{n-1}{k-1}}{\binom{n-2}{n-k-1} + \binom{n-1}{n-k-1}} = \frac{k(2n-k-1)}{(n-k)(n+k-1)}. \tag{1.2}$$

Mills, Robbins, and Rumsey had no explanation for this pattern in the ratios, but they were able to confirm that it persists at least through the first twenty rows. If this conjecture is correct, then the value of each $A_{n,k}$ is uniquely determined by $A_{n,k-1}$when $k > 1$ and by $A_{n,1} = \sum_{k=1}^{n-1} A_{n-1,k}$. Conjecture 1 is equivalent to the following explicit formula for $A_{n,k}$.

Conjecture 2 *For* $1 \le k \le n$,

$$A_{n,k} = \binom{n+k-2}{k-1} \frac{(2n-k-1)!}{(n-k)!} \prod_{j=0}^{n-2} \frac{(3j+1)!}{(n+j)!}. \tag{1.3}$$

The equivalence of these two conjectures will be proven in Chapter 5 when we investigate hypergeometric series. Conjecture 2 implies a formula for the number of $n \times n$ alternating sign matrices.

Conjecture 3 (The ASM conjecture) *The total number of* $n \times n$ *alternating sign matrices is*

$$A_n = A_{n+1,1} = \prod_{j=0}^{n-1} \frac{(3j+1)!}{(n+j)!}. \tag{1.4}$$

Mills, Robbins, and Rumsey were unable to prove any of these conjectures.

Exercises

1.1.1 Show that if V is the column vector with entries $(1, 2, 3, \ldots, n)$ and if S is the matrix that corresponds to the permutation σ, then SV is the column vector with entries $(\sigma(1), \sigma(2), \ldots, \sigma(n))$.

1.1.2 Prove that the product of two permutation matrices of the same size must be a permutation matrix.

1.1.3 Prove that if S is the permutation matrix for σ and T is the permutation matrix for τ, then $T \cdot S$ is the permutation matrix for $\sigma \circ \tau$.

1.1.4 Show by example that the product of two alternating sign matrices does not have to be an alternating sign matrix.

1.1.5 Find the largest subset of the set of $n \times n$ alternating sign matrices that contains the permutation matrices and is closed under matrix multiplication.

1.1.6 The order of a function, $f(n)$, is the greatest lower bound of the set of α for which $\lim_{n\to\infty}[\log|f(n)|]/n^{\alpha} = 0$. Show that Stirling's formula: $\log n! \approx n \log n - n$ and Conjecture 3 imply that A_n has order 2.

1.1.7 Verify that Conjecture 2 implies both Conjecture 1 and Conjecture 3.

1.1.8 Check the simplification in equation (1.2).

1.1.9 Verify that $c_{n,k} = \binom{n-2}{k-1} + \binom{n-1}{k-1}$ satisfies

1. $c_{n,1} = 2$,
2. $c_{n,n-1} = n$,
3. $c_{n,k} = c_{n-1,k-1} + c_{n-1,k}$.

1.1.10 Explain why each alternating sign matrix has exactly one 1 in the top row.

1.1.11 What does the Pascal-like triangle look like if we separate *permutation* matrices according to the position of the 1 in the first row?

1.1.12 The *Mathematica* code given below allows you to calculate the numbers `A[n,k]` that satisfy Conjecture 1. An n-line table of values is produced by the command `ASMtable[n]`. What conjecture can you make about the parity of these values?

```
A[1,1]=1;
A[n_,1]:=A[n,1]=Sum[A[n-1,k],{k,n-1}];
A[n_,k_]:=A[n,k]
          =((n-k+1)(n+k-2)/((k-1)(2n-k))) A[n,k-1];
ASMtable[m_]
     :=TableForm[ Table[Table[A[n,k],{k,n}],{n,m}]];
```

1.1.13 Prove the parity conjecture made in exercise 1.1.12.

1.1.14 Prove that $\prod_{j=0}^{n-1}(3j+1)!/(n+j)!$ is an integer for each positive integer n.

Figure 1.2. George E. Andrews.

1.2 Connections to plane partitions

When you have a conjecture that you cannot prove, you turn to others for help. This was the situation in which Mills, Robbins, and Rumsey soon found themselves.

> After several months of trying to prove our conjecture about the number of alternating sign matrices, we began to suspect that the theory of plane partitions might be involved. We called Richard Stanley and told him about our conjecture. He startled us a few days later when he told us that, although he did not know a proof of our conjecture, he had seen the sequence before, and that it was known to enumerate another class of combinatorial objects which had appeared only a few years earlier in a paper by George Andrews. Naturally we headed straight for the library to find out what it was all about. (Robbins 1991)

George Andrews (1979) had been counting descending plane partitions.

We can visualize a **plane partition** as stacks of unit cubes pushed into a corner as in Figure 1.3. In this example, there are 75 cubes, and we refer to it as a plane partition of 75. There are a lot of ways of stacking cubes into a corner. If we let $pp(n)$ denote the number of plane partitions of n, then

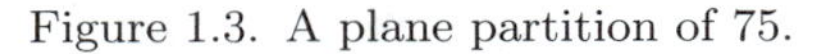

Figure 1.3. A plane partition of 75.

n	$pp(n)$
1	1
2	3
3	6
4	13
5	24
6	48
7	86
8	160
9	282
10	500
$\vdots$	$\vdots$
75	37,745,732,248,153.

Actually looking for all possible arrangements and counting each one is impractical except for very small values of n. To find the exact value of $pp(n)$, we can use a generating function.

The **generating function** for plane partitions is the power series whose coefficients are given by $pp(n)$:

$$pp(0) + pp(1)\, q + pp(2)\, q^2 + pp(3)\, q^3 + \cdots .$$

To quote Herb Wilf (1994), "A generating function is a clothesline on which we hang up a sequence of numbers for display." The choice of q as the variable, rather than x, has historical roots that relate to a connection to Fourier series and Jacobi's theta functions where q is often used as shorthand for $e^{2\pi i\tau}$. It is convenient to define $pp(0) = 1$. The one plane partition of 0 is the empty plane partition with no cubes.

The following theorem was discovered by Major Percy Alexander MacMahon (1897), although he did not publish a proof until 1912. We shall see one proof in Chapter 3 and a different proof in Chapter 4.

Theorem 1.1 *The generating function for plane partitions can be expressed as*

$$\begin{aligned}\sum_{n=0}^{\infty} pp(n)\, q^n &= \frac{1}{(1-q)(1-q^2)^2(1-q^3)^3(1-q^4)^4\cdots} \\ &= \prod_{j=1}^{\infty} \frac{1}{(1-q^j)^j}. \qquad (1.5)\end{aligned}$$

Equation (1.5) is valid for $|q| < 1$, but we can also think of our power series as a **formal power series** in which we treat it as no more than a clothesline that satisfies certain algebraic relations that are encoded by the product side. Differentiation is among the legal manipulations that we are allowed to perform. If we differentiate each side of this equation, we obtain the following recursive relationship for $pp(n)$ which we shall prove in the next chapter.

Theorem 1.2 *If $pp(n)$ is the number of plane partitions of n and $\sigma_2(n)$ is the sum of the squares of the divisors of n, then*

$$pp(n) = \frac{1}{n} \sum_{j=1}^{n} \sigma_2(j)\, pp(n-j). \qquad (1.6)$$

As an example, $\sigma_2(6) = 1^2 + 2^2 + 3^2 + 6^2 = 50$. The partial table given below shows the values of $\sigma_2(n)$ up to $n = 10$ and the values of $pp(n)$ up to $n = 5$:

n	0	1	2	3	4	5	6	7	8	9	10
$\sigma_2(n)$		1	5	10	21	26	50	50	85	91	130
$pp(n)$	1	1	3	6	13	24					

Figure 1.4. Percy A. MacMahon (*left*) and Ian G. Macdonald.

We then have that

$$\begin{aligned} pp(6) &= \frac{1}{6}(1\cdot 24 + 5\cdot 13 + 10\cdot 6 + 21\cdot 3 + 26\cdot 1 + 50\cdot 1) \\ &= 288/6 \;=\; 48. \end{aligned}$$

Restricted plane partitions

Plane partitions appeared for the first time in a paper by Major MacMahon presented to the Royal Society in 1896 and published in 1897. Percy A. MacMahon was born on Malta in 1854, the son of a brigadier-general. He attended the Royal Military Academy in Woolwich, became an artillery officer, served for five years in India, and saw action on the Northwest Frontier. In 1882, he returned to the Royal Military Academy, first as a mathematics instructor and later as Professor of Physics. He retired from the army in 1898 and worked until 1920 as Deputy Warden of Standards for the Board of Trade. From 1904 until his death in 1929 he was an honorary member of St. John's College, Cambridge, and spent much time there. His two-volume *Combinatory Analysis* (1915–1916) helped establish combinatorics as a discipline in its own right.

Figure 1.5. A symmetric plane partition.

From here on, it will be convenient to describe a plane partition by the positions of the unit cubes. We define a **plane partition** to be a finite subset, $\mathcal{P}$, of positive integer lattice points, $\{(i,j,k)\} \subset \mathbf{N}^3$, such that if (r,s,t) lies in $\mathcal{P}$ and if (i,j,k) satisfies $1 \leq i \leq r$, $1 \leq j \leq s$, and $1 \leq k \leq t$, then (i,j,k) also lies in $\mathcal{P}$. We define

$$\mathcal{B}(r,s,t) = \{(i,j,k) \mid 1 \leq i \leq r,\ 1 \leq j \leq s,\ 1 \leq k \leq t\}.$$

In Chapter 3 and again, using a different approach, in Chapter 4 we shall prove Theorem 1.1 as a special case of MacMahon's generating function for plane partitions contained in $\mathcal{B}(r,s,t)$.

Theorem 1.3 *The generating function for plane partitions that are subsets of* $\mathcal{B}(r,s,t)$ *is given by*

$$\prod_{i=1}^{r}\prod_{j=1}^{s}\frac{1-q^{i+j+t-1}}{1-q^{i+j-1}} \tag{1.7}$$

Long before he proved this theorem, MacMahon was already conjecturing other generating functions. One of his most notable conjectures – one that plays an important role in our story – was an identity for the generating function for **symmetric plane partitions**, plane partitions for which (i,j,k) is an element if and only if (j,i,k) is an element. These are plane partitions with symmetry about the $x=y$ plane as shown in Figure 1.5.

In the 1970s, Ian Macdonald recognized that the generating function given in Theorem 1.3 can also be expressed as

$$\prod_{(i,j,k)\in\mathcal{B}(r,s,t)} \frac{1-q^{i+j+k-1}}{1-q^{i+j+k-2}}.$$

This led him to an appealing way of expressing MacMahon's conjecture for symmetric plane partitions. We let $\mathcal{S}_2$ denote the group of permutations acting on the first two coordinates of (i,j,k). This group has two elements: the identity which leaves the triple unchanged and the operation which interchanges the first two coordinates. We let $\mathcal{B}/\mathcal{S}_2$ denote the set of **orbits** of elements of $\mathcal{B}$ under the action of $\mathcal{S}_2$. Two elements are in the same orbit if and only if there is an element of $\mathcal{S}_2$ that transforms one into the other. In this case, there are only two types of orbits: singletons, (i,i,k), and doubletons, (i,j,k) and (j,i,k) where $i \neq j$. If η is an orbit, then we use $|\eta|$ to denote the number of elements in η.

The **height** of an element (i,j,k) is defined as the number

$$\mathrm{ht}(i,j,k) = i+j+k-2.$$

The height of the corner position, $(1,1,1)$, is 1, and the height increases by one for each step away from the corner. The **height of an orbit**, $\mathrm{ht}(\eta)$, is the height of any one of the elements in that orbit.

Conjecture 4 (The MacMahon conjecture) *The generating function for symmetric plane partitions that are subsets of $\mathcal{B}(r,r,t)$ is given by*

$$\prod_{\eta\in\mathcal{B}(r,r,t)/\mathcal{S}_2} \frac{1-q^{|\eta|(1+\mathrm{ht}(\eta))}}{1-q^{|\eta|\mathrm{ht}(\eta)}}. \tag{1.8}$$

This particular formulation of the conjecture is due to Macdonald. MacMahon expressed this generating function as

$$\left(\prod_{i=1}^{r}\prod_{k=1}^{t} \frac{1-q^{1+2i+k-2}}{1-q^{2i+k-2}}\right)\left(\prod_{1\le i<j\le r}\prod_{k=1}^{t} \frac{1-q^{2+2(i+j+k-2)}}{1-q^{2(i+j+k-2)}}\right).$$

The first piece is the product over singleton orbits while the second is the product over doubleton orbits.

This is no longer a conjecture. It was independently proven in the late 1970s by two mathematicians who took very different approaches: George Andrews (1978) and Ian Macdonald (1979). We shall see a version of Macdonald's proof in Chapter 4.

Once MacMahon's conjecture is seen in this format, other possible conjectures spring to mind. We can try to replace $\mathcal{S}_2$ with other groups of permutations on the coordinates (i, j, k). Theorem 1.3 is what you get if you replace $\mathcal{S}_2$ with the trivial group that is made up of just the identity. In our new notation, the generating function in Theorem 1.3 is written as

$$\prod_{\eta \in \mathcal{B}(r,s,t)} \frac{1 - q^{1+\mathrm{ht}(\eta)}}{1 - q^{\mathrm{ht}(\eta)}}.$$

There are two other groups that act on these three coordinates: $\mathcal{S}_3$, the group of all permutations, and $\mathcal{C}_3$, the group of **cyclic permutations** which consists of

$$(i, j, k) \longmapsto (i, j, k), \quad (i, j, k) \longmapsto (j, k, i), \quad \text{and} \quad (i, j, k) \longmapsto (k, i, j).$$

MacMahon had tried to guess the generating function for plane partitions that are invariant under all permutations of the vertices, what are now called **totally symmetric plane partitions** in which (i, j, k) is an element if and only if each permutation of these vertices yields an element of the plane partition. The obvious guess,

$$\prod_{\eta \in \mathcal{B}(r,r,r)/\mathcal{S}_3} \frac{1 - q^{|\eta|(1+\mathrm{ht}(\eta))}}{1 - q^{|\eta|\mathrm{ht}(\eta)}}, \tag{1.9}$$

is wrong. If we simplify this product with r as small as 3, we find that it is not a polynomial. But MacMahon missed two conjectures that Macdonald spotted. The first is that while the product in (1.9) is not a polynomial, if it *were* the generating function for totally symmetric plane partitions that are subsets of $\mathcal{B}(r, r, r)$, then setting $q = 1$ would give the total number of distinct totally symmetric plane partitions that are subsets of $\mathcal{B}(r, r, r)$ (including the empty plane partition). Surprisingly, even though the generating function is wrong, taking the limit as q approaches 1 appeared to yield the right number.

Conjecture 5 (The TSPP conjecture) *The total number of totally symmetric plane partitions that are subsets of* $\mathcal{B}(r, r, r)$ *is*

$$\prod_{\eta \in \mathcal{B}(r,r,r)/\mathcal{S}_3} \frac{1 + \mathrm{ht}(\eta)}{\mathrm{ht}(\eta)}. \tag{1.10}$$

The second conjecture that MacMahon missed is the generating function for **cyclically symmetric plane partitions**, plane partitions for which (i, j, k) is an element if and only if all cyclic permutations of these vertices are also elements.

Conjecture 6 (The Macdonald conjecture) *The generating function for cyclically symmetric plane partitions that are subsets of $\mathcal{B}(r, r, r)$ is*

$$\prod_{\eta \in \mathcal{B}(r,r,r)/\mathcal{C}_3} \frac{1 - q^{|\eta|(1+\mathrm{ht}(\eta))}}{1 - q^{|\eta|\,\mathrm{ht}(\eta)}}. \tag{1.11}$$

Conjecture 5 was proven by John Stembridge (1995). In the late 1970s and early '80s there was a great deal of effort expended on trying to prove Conjecture 6. Richard Stanley (1981) wrote about it:

If I had to single out the most interesting open problem in all of enumerative combinatorics, this would be it.

Here is where Mills, Robbins, and Rumsey re-enter our story, for before the spring of 1982 they had succeeded in proving Conjecture 6 by trying to understand how they might prove their conjectures for alternating sign matrices.

Exercises

1.2.1 Given below is the *Mathematica* code to calculate $pp(n)$ using the recursion of Theorem 1.2. The function `ppTable(`N`)` produces a table of values of $pp(n)$ up to $n = N$. What fraction of the values of $pp(n)$ up to $n = 100$ are even, are multiples of 3, 5, or 7? Are these numbers consistent with the hypothesis that the values of $pp(n)$ are randomly distributed with respect to the moduli 2, 3, 5, and 7?

```
pp[0]=1;
pp[n_]:=pp[n]
      =Sum[ DivisorSigma[2,j] pp[n-j],{j,n}]/n;
ppTable[n_]:=TableForm[ Table[{j,pp[j]},{j,0,n}]];
```

1.2.2 Show that the two product formulæ for the generating function for plane partitions in $\mathcal{B}(r, s, t)$ are equivalent:

$$\prod_{i=1}^{r}\prod_{j=1}^{s} \frac{1 - q^{i+j+t-1}}{1 - q^{i+j-1}} = \prod_{(i,j,k)\in\mathcal{B}(r,s,t)} \frac{1 - q^{i+j+k-1}}{1 - q^{i+j+k-2}}.$$

1.2.3 The *Mathematica* code given below finds the generating function for plane partitions in $\mathcal{B}(r,s,t)$. How many plane partitions of 37 fit inside $\mathcal{B}(5,5,5)$? How does this compare with the total number of plane partitions of 37? What is the total number of plane partitions that can be fit inside $\mathcal{B}(5,5,5)$?

```
PP[r_,s_,t_,q_]:= Expand[Factor[
     Product[ (1-q^(i+j+t-1))/(1-q^(i+j-1)),
     {i,r},{j,s}]]]
```

1.2.4 Prove that there are rt singleton orbits in $\mathcal{B}(r,r,t)/\mathcal{S}_2$, and $\binom{r}{2}t$ doubleton orbits.

1.2.5 How many orbits of size 1, 2, 3, or 6 are there in $\mathcal{B}(r,r,r)/\mathcal{S}_3$?

1.2.6 How many orbits of size 1, 2, 3, or 6 are there in $\mathcal{B}(r,r,r)/\mathcal{C}_3$?

1.2.7 Show that the generating function for symmetric plane partitions in $\mathcal{B}(3,3,3)/\mathcal{S}_2$ simplifies to

$$\frac{(1-q^{10})(1-q^{12})(1-q^{14})}{(1-q)(1-q^3)(1-q^5)}.$$

1.2.8 Show that the generating function for symmetric plane partitions in $\mathcal{B}(r,r,t)/\mathcal{S}_2$ simplifies to

$$\prod_{i=1}^{r}\frac{1-q^{2i+t-1}}{1-q^{2i-1}}\prod_{1\le i<j\le r}\frac{1-q^{2i+2j+2t-2}}{1-q^{2i+2j-2}}.$$

1.2.9 The *Mathematica* code given below generates $\mathtt{SPP}(r,t,q)$, the generating function for symmetric plane partitions in $\mathcal{B}(r,r,t)$. How many symmetric plane partitions of 37 lie inside $\mathcal{B}(5,5,5)$? What is the total number of symmetric plane partitions that lie inside $\mathcal{B}(5,5,5)$?

```
SPP[r_,t_,q_]:= Expand[Factor[
     Product[(1-q^(2i+t-1))/(1-q^(2i-1)),{i,r}]
     Product[ (1-q^(2i+2j+2t-2))/(1-q^(2i+2j-2)),
     {i,r-1},{j,i+1,r}]]]
```

1.2.10 Verify that

$$\prod_{\eta\in\mathcal{B}(3,3,3)/\mathcal{S}_3}\frac{1-q^{|\eta|(1+\mathrm{ht}(\eta))}}{1-q^{|\eta|\,\mathrm{ht}(\eta)}}$$

is not a polynomial. Why does the generating function for totally symmetric plane partitions inside $\mathcal{B}(3,3,3)$ have to be a polynomial?

1.2.11 Prove that Conjecture 5 is equivalent to the statement that the number of totally symmetric plane partitions inside $\mathcal{B}(r,r,r)$ satisfies the recursive formula

$$\begin{aligned}\texttt{TSPP}(0) &= 1,\\ \texttt{TSPP}(r) &= \prod_{i=1}^{r} \frac{i+2r-1}{2i+r-2}\,\texttt{TSPP}(r-1).\end{aligned}$$

1.2.12 Find the five totally symmetric plane partitions that fit inside $\mathcal{B}(2,2,2)$.

1.2.13 How many totally symmetric plane partitions fit in $\mathcal{B}(5,5,5)$?

1.2.14 Prove that Conjecture 6 is equivalent to the statement that the generating function for cyclically symmetric plane partitions inside $\mathcal{B}(r,r,r)$ satisfies the recursive formula

$$\begin{aligned}\texttt{CSPP}(0,q) &= 1,\\ \texttt{CSPP}(r,q) &= \frac{1-q^{3r-1}}{1-q^{3r-2}} \prod_{i=1}^{r-1} \frac{1-q^{6r+3i-3}}{1-q^{3r+3i-3}}\,\texttt{CSPP}(r-1,q).\end{aligned}$$

1.2.15 The *Mathematica* code given below produces the generating function for cyclically symmetric plane partitions in $\mathcal{B}(r,r,r)$. How many cyclically symmetric plane partitions of 37 lie inside $\mathcal{B}(5,5,5)$? What is the total number of cyclically symmetric plane partitions that fit inside $\mathcal{B}(5,5,5)$?

```
CSPP[0,q_]:=1;
CSPP[r_,q_]:=CSPP[r,q]=Expand[Factor[
  (1-q^(3r-1))/(1-q^(3r-2))*
  Product[ (1-q^(6r+3i-3))/(1-q^(3r+3i-3)),{i,r-1}]
  *CSPP[r-1,q]]]
```

1.3 Descending plane partitions

There is a general technique that we shall see in Chapter 3 that enables us to express the generating functions for many different types of plane partitions as determinants of matrices with polynomial entries. For cyclically symmetric plane partitions that fit inside $\mathcal{B}(r,r,r)$, the generating function is given by the determinant of an $r \times r$ matrix:

$$\det\left(\delta_{ij} + q^{3i-2}\begin{bmatrix} i+j-2 \\ j-1 \end{bmatrix}_{q^3}\right)_{i,j=1}^{r}, \qquad (1.12)$$

Figure 1.6. Separating a cyclically symmetric plane partition into shells.

where δ_{ij} is 1 if $i = j$, and it is 0 otherwise, and

$$\begin{bmatrix} a \\ j \end{bmatrix}_q = \prod_{i=1}^{j} \frac{1 - q^{a-j+i}}{1 - q^i} = \frac{(1 - q^a)(1 - q^{a-1}) \cdots (1 - q^{a-j+1})}{(1 - q)(1 - q^2) \cdots (1 - q^j)}.$$

The empty product $(j = 0)$ is defined to be 1. This rational function of q is actually a polynomial when a is a non-negative integer. We shall study it in detail in Section 3.1.

The key to constructing the generating function for cyclically symmetric plane partitions is to separate the plane partition into **shells**. The first or outermost shell consists of those positions in the plane partition for which at least one of the coordinates is 1. The second shell is the subset of points which are not in the first shell and have at least one coordinate equal to 2. In general, the rth shell consists of those elements of the plane partition for which the minimal coordinate value is r (see Fig. 1.6).

Each shell must have cyclic symmetry, and this simply means that each of the three sides of a shell is a copy of the others. In other words, each shell is uniquely determined by one of its sides (see Fig. 1.7). We can thus uniquely reconstruct our cyclically symmetric plane partition if we just keep the bottom level of each shell. We can now encode our symmetric plane partition by describing each level as a weakly decreasing sequence of positive integers, each integer recording the length of one row

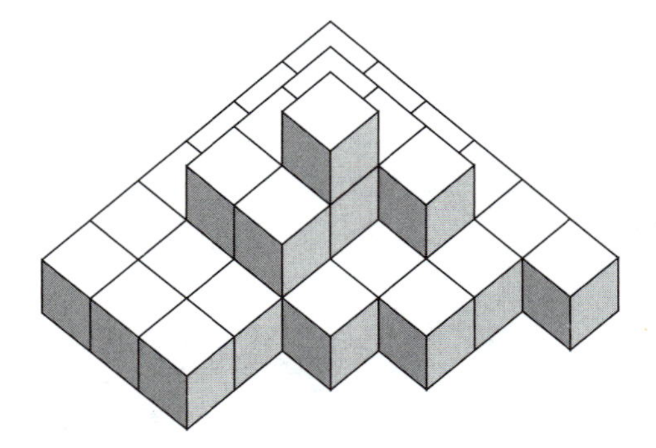

Figure 1.7. Removing all but one side of each shell and then restacking.

in that level. The cyclically symmetric plane partition that we have been using as an example encodes as

$$\begin{array}{cccccc} 6 & 5 & 5 & 4 & 3 & 3 \\ & 3 & 2 & 2 & & \\ & & 1 & & & \end{array}.$$

Because we need to be able to construct a symmetric shell from each level, the number of integers in each sequence must equal the largest integer in the sequence. By indenting each sequence by one, the condition that each shell must fit inside the previous shell is equivalent to the condition that each integer must be strictly less than the integer directly above it. We call such an arrangement of positive integers with each row indented, weak decrease across rows, and strict decrease down columns a **strict shifted plane partition**. We can view cyclically symmetric plane partitions as a subset of the strict shifted plane partitions.

Andrews was unable to evaluate the determinant given in (1.12), so he went after an easier problem. If Macdonald's conjecture were true, then it would also be true in the limit as q approaches 1:

$$\det\left(\delta_{ij} + \binom{i+j-2}{j-1}\right)_{i,j=1}^{r} = \prod_{\eta \in \mathcal{B}(r,r,r)/\mathcal{C}_3} \frac{|\eta|(1+\mathrm{ht}(\eta))}{|\eta|\,\mathrm{ht}(\eta)}. \tag{1.13}$$

Andrews was able to prove this, though it took him 33 pages and some heavy transformation formulæ for hypergeometric series. The general case looked unassailable, at least from this vantage.

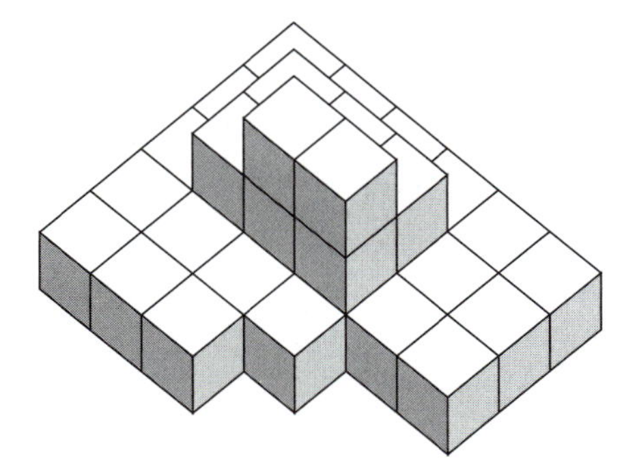

Figure 1.8. A descending plane partition.

Something extra

Though Andrews had fallen short of proving Macdonald's conjecture, he had developed some powerful techniques for attacking it and had come much closer than anyone else. As he realized, these techniques were applicable to other problems.

Andrews had counted the total number of strict shifted plane partitions for which the largest part in each row equals the number of parts in that row. A modification of his argument allowed him to find the generating function for what he would call **descending plane partitions**. These are strict shifted plane partitions in which the number of parts in each row is strictly less than the largest part in that row and is greater than or equal to the largest part in the next row (Fig. 1.8). As we shall see in Chapter 3, the generating function for descending plane partitions in which the largest part is less than or equal to r can be written in matrix form as

$$\det \left(\delta_{ij} + q^{i+1} \begin{bmatrix} i+j \\ j-1 \end{bmatrix}_q \right)_{i,j=1}^{r-1}.$$

Andrews conjectured that this generating function could be written as a product.

Conjecture 7 (The Andrews conjecture) *The generating function for descending plane partitions with largest part less than or equal to r*

is

$$\det\left(\delta_{ij} + q^{i+1}\begin{bmatrix} i+j \\ j-1 \end{bmatrix}_q\right)_{i,j=1}^{r-1} = \prod_{1\le i\le j\le r} \frac{1-q^{r+i+j-1}}{1-q^{2i+j-1}}. \qquad (1.14)$$

Again, he was unable to evaluate this determinant, but the tools that had yielded the $q = 1$ case of Macdonald's conjecture worked on his own conjecture. He was able to prove that

$$\det\left(\delta_{ij} + \binom{i+j}{j-1}\right)_{i,j=1}^{r-1} = \prod_{1\le i\le j\le r} \frac{r+i+j-1}{2i+j-1}. \qquad (1.15)$$

Richard Stanley received this result in late 1978 or early 1979, roughly two years before he heard from Mills, Robbins, and Rumsey about their counting problem. He had undoubtedly worked out the first few values for the number of descending plane partitions: 1, 2, 7, 42, 429, 7435, ..., because almost immediately he recognized the counting sequence for alternating sign matrices. I leave it as an exercise for the reader to verify that the two formulæ are in fact identical:

$$\prod_{1\le i\le j\le r} \frac{r+i+j-1}{2i+j-1} = \prod_{j=0}^{r-1} \frac{(3j+1)!}{(r+j)!}. \qquad (1.16)$$

Andrews had proven that this product counts the total number of descending plane partitions with largest part less than or equal to r. If Mills, Robbins, and Rumsey could find a one-to-one correspondence between the alternating sign matrices of size r and the descending plane partitions with largest part less than or equal to r, then they would have a proof of Conjecture 3. The mathematical community is still looking for this correspondence, but the directions in which it led Mills, Robbins, and Rumsey were extremely fruitful. They were to discover a proof of Conjecture 7, and from there they were able to prove Conjecture 6.

The moral

The insights that Mills, Robbins, and Rumsey were able to obtain would not have been possible if Andrews had not decided to use his technique on a problem with absolutely no relevance other than the fact that his technique would work on it. The only hint that this determinant evaluation would prove useful is the fact that its product formula is surprisingly simple, that the solution has what a mathematician recognizes as *elegance*.

Looking for symmetries

Mills, Robbins, and Rumsey were looking for a correspondence between alternating sign matrices of size r and descending plane partitions with largest part less than or equal to r. It soon became obvious that such a correspondence could not be simple. To illustrate some of the problems, we list the seven alternating sign matrices of size 3:

$$\begin{pmatrix} 1 & 0 & 0 \\ 0 & 1 & 0 \\ 0 & 0 & 1 \end{pmatrix}, \quad \begin{pmatrix} 1 & 0 & 0 \\ 0 & 0 & 1 \\ 0 & 1 & 0 \end{pmatrix},$$

$$\begin{pmatrix} 0 & 1 & 0 \\ 1 & 0 & 0 \\ 0 & 0 & 1 \end{pmatrix}, \quad \begin{pmatrix} 0 & 1 & 0 \\ 1 & -1 & 1 \\ 0 & 1 & 0 \end{pmatrix}, \quad \begin{pmatrix} 0 & 1 & 0 \\ 0 & 0 & 1 \\ 1 & 0 & 0 \end{pmatrix},$$

$$\begin{pmatrix} 0 & 0 & 1 \\ 1 & 0 & 0 \\ 0 & 1 & 0 \end{pmatrix}, \quad \begin{pmatrix} 0 & 0 & 1 \\ 0 & 1 & 0 \\ 1 & 0 & 0 \end{pmatrix}.$$

and the seven descending plane partitions with largest part less than or equal to three ($\emptyset$ denotes the empty partition with no parts):

$$\emptyset \qquad 2 \qquad 3 \qquad 3\ 1 \qquad 3\ 2 \qquad 3\ 3 \qquad \begin{array}{cc} 3 & 3 \\ & 2 \end{array}.$$

The alternating sign matrices are rich in symmetries, but there is little apparent symmetry among the descending plane partitions. If we really have a natural correspondence between these two sets of objects, then descending plane partitions must have hidden symmetries. The most promising path toward a natural correspondence starts with guesses about what these hidden symmetries might be.

We have another landmark. We know that the position of the 1 in the first row of the alternating sign matrix is an important parameter. It was in keeping track of this parameter that Robbins and Rumsey were able to discover the pattern that led to their conjecture for the number of alternating sign matrices. If there is a natural correspondence between alternating sign matrices and descending plane partitions, then there must be something in a descending plane partition that corresponds to the position of the 1 in the first row. For the descending plane partitions with largest part less than or equal to three, we need a natural way of separating this set into disjoint subsets of cardinalities two, three, and two. Mills, Robbins, and Rumsey made a guess. They subdivided the set of descending plane partitions with largest part less than or equal to r into subsets according to the number of parts of size r.

Conjecture 8 *The number of descending plane partitions with largest part less than or equal to r and for which r appears as a part exactly $k-1$ times is equal to the number of $r \times r$ alternating sign matrices with a 1 in the kth column of the first row.*

Combining this with Conjecture 2, they made the following assertion.

Conjecture 9 *The number of descending plane partitions with largest part less than or equal to r and for which r appears as a part exactly $k-1$ times is equal to*

$$\binom{r+k-2}{k-1} \frac{(r-k-1)!}{(r-k)!} \prod_{j=0}^{r-2} \frac{(3j+1)!}{(r+j)!}.$$

Several things served to elevate their guess to the status of a conjecture. First, they confirmed it numerically as far as $r = 10$. Second, they observed special cases in which they could prove it to be true. We know that if the 1 in the first row of the alternating sign matrix occurs in the first or last column, then we are simply counting alternating sign matrices of the next size down. If a descending plane partition has no parts of size r or $r-1$ parts of size r (and thus the bottom level is filled with the maximal number of cubes allowed), then we are counting descending plane partitions with largest part less than or equal to $r-1$. Third, this conjecture explains one of the hidden symmetries. We know that

$$A_{r,k} = A_{r,r-k+1}.$$

If Conjecture 8 is correct, then the number of descending plane partitions with $k-1$ parts of size r is equal to the number of descending plane partitions with $r-k$ parts of size r. Fourth, Conjecture 8 implies Conjecture 9 which, at the time it was made, had no other supporting evidence. When Conjecture 9 was proven to be true, everyone was convinced that Conjecture 8 had to be true.

Surprise!

Mills, Robbins, and Rumsey wanted to prove Conjecture 8. They could not. But they did prove Conjecture 9. Their proof employed standard techniques of linear algebra, a very simple summation formula for hypergeometric series, and a good deal of ingenuity. Conjecture 9 implies Andrews's result for the total number of descending plane partitions, equation (1.15). The simplicity of their proof suggested it might do more. It did. They were able to prove Andrews's Conjecture 7 for the generating function for descending plane partitions. In fact, they were able to find the generating function for descending plane partitions with largest part less than or equal to r and exactly $k-1$ parts of size r.

This emboldened them to look at Macdonald's Conjecture 6. The key to proving Andrews's conjecture had been the recognition that the number of parts of maximal allowed size was a critically important parameter. They tried a parallel approach to the generating function for cyclically symmetric plane partitions. It was harder, but it worked. We shall see this proof in Chapter 5.

In retrospect, their success is not totally surprising. It is often the case that the key to a good proof – especially an inductive proof as this turned out to be – lies in knowing which additional parameters to introduce. In recognizing the connection between descending plane partitions and alternating sign matrices, they saw that the number of appearances of parts of size r was such a parameter.

More than the numerical evidence for the refined ASM conjecture (Conjecture 1) or even its aesthetic appeal, this application of its fundamental insight to an important unsolved problem convinced everyone that the conjecture of Mills, Robbins, and Rumsey had to be true, though it gave not the slightest hint of why it should be true or how it could be proven.

A happy ending

Doron Zeilberger announced a proof of Conjecture 3 in December of 1992, and in the fall of 1995 he discovered a proof of Conjectures 1 and 2. The story of the passage from the early work of Mills, Robbins, and Rumsey to the ultimate triumph of Zeilberger will be told in Chapter 6. It is a tale of many explorers who fanned out across the lands that Mills, Robbins, and Rumsey had discovered. The final breakthrough came from Greg Kuperberg (1996b), who discovered the tracks of previous explorers: Physicists – specifically, those working in statistical mechanics – had been here first. They had been studying alternating sign matrices for years, only they had called them "square ice."

```
 H—O     H—O—H     O—H     O—H     O—H
   |                 |       |       |
   H       H         H       H       H
           |
 H—O—H     O     H—O     H—O—H     O—H
           |       |                 |
   H       H       H       H         H
   |                       |
 H—O     H—O—H     O—H     O     H—O—H
                   |       |
   H       H       H       H         H
   |       |                         |
 H—O     H—O     H—O     H—O—H     O—H
                   |
   H       H       H       H         H
   |       |               |         |
 H—O     H—O     H—O—H     O—H     O—H
```

Figure 1.9. Square ice.

Square ice is a two-dimensional arrangement of water molecules, H_2O, with oxygen atoms at the vertices of a square lattice and one hydrogen atom between each pair of adjacent oxygen atoms (see Fig. 1.9). Each molecule corresponds to an entry in an alternating sign matrix: horizontal molecules are 1s, vertical molecules are -1s, and all other molecules are 0s. In order to get a one-to-one correspondence with alternating sign matrices, we must restrict our sheets of square ice to those with no hydrogen atoms above the top row or below the bottom row. The pattern of square ice in this example corresponds to the alternating sign matrix

$$\begin{pmatrix} 0 & 1 & 0 & 0 & 0 \\ 1 & -1 & 0 & 1 & 0 \\ 0 & 1 & 0 & -1 & 1 \\ 0 & 0 & 0 & 1 & 0 \\ 0 & 0 & 1 & 0 & 0 \end{pmatrix}.$$

Physicists are interested in the weighted sums of all possible configurations occurring in vast sheets of square ice, but Kuperberg realized that their methods could be applied to the modest counting problem that the mathematicians wished to solve. As we shall see in Chapter 7, while the physicists had not solved the problem of counting alternating sign matrices, they had developed the tools that were needed for the final ascent.

Summary of the fourteen conjectures

The fourteen conjectures around which this story is woven are listed below for reference.

1. The formula for $A_{n.k}/A_{n,k+1}$. Page 7. Posed in Mills, Robbins, Rumsey (1983). Proven in Zeilberger (1996b). Proof given in Chapter 7.
2. The formula for $A_{n,k}$. Page 7. Posed in Mills, Robbins, Rumsey (1983). As shown in Section 5.2, this is equivalent to Conjecture 1 and thus follows from the proof in Chapter 7.
3. The formula for A_n. Page 7. Posed in Mills, Robbins, Rumsey (1983). First proven in Zeilberger (1996a). Next proven in Kuperberg (1996b). This is a special case of Conjecture 2 and thus also follows from the proof in Chapter 7.
4. The generating function for symmetric plane partitions. Page 14. Posed in MacMahon (1899). Independently proven in Andrews (1978) and Macdonald (1979). Proof given in Section 4.3.
5. The number of totally symmetric plane partitions. Page 15. Posed by Macdonald. Proven in Stembridge (1995). Proof not in this book.
6. The generating function for cyclically symmetric plane partitions. Page 16. Posed in Macdonald (1979). Proven in Mills, Robbins, Rumsey (1982) as a special case of a general result that also implies Conjectures 7 and 9. Proof given in Section 5.3.
7. The generating function for descending plane partitions. Page 21. Posed in Andrews (1979). Proven in Mills, Robbins, Rumsey (1982) as a special case of a general result that also implies Conjectures 6 and 9. Proof given in Section 5.3.
8. The descending plane partition–ASM correspondence. Page 24. Shown in Mills, Robbins, Rumsey (1982) to be equivalent to Conjectures 1 and 2 and thus follows from the proof in Chapter 7.
9. The descending plane partition formula. Page 24. Posed and proven in Mills, Robbins, Rumsey (1982) as a special case of a general result that also implies Conjectures 6 and 7. Proof given in Section 5.3.

10. The refined descending plane partition–ASM correspondence. Page 193. Posed in Mills, Robbins, Rumsey (1983). This conjecture is still unproven. It implies Conjecture 8 and thus also Conjectures 1 and 2.
11. The 3-count of ASMs. Page 193. Posed in Mills, Robbins, Rumsey (1986). Proven in Kuperberg (1996b) as a byproduct of his proof of Conjecture 3. Proof not given in this book.
12. The TSSCPP–ASM correspondence. Page 195. Posed in Mills, Robbins, Rumsey (1986). Shown in Andrews (1994) to be equivalent to Conjecture 3. Follows from the proof in Chapter 7.
13. The orbit counting generating function for TSPPs. Page 200. Independently posed by Andrews and Robbins. This conjecture is still unproven. It implies Conjecture 5.
14. The gog–magog correspondence. Page 218. Posed in Mills, Robbins, Rumsey (1986). Proven in Zeilberger (1996a). As a result of Andrews (1994), this conjecture implies Conjectures 3 and 12. Proof not given in this book.

Exercises

1.3.1 For each of the following strict shifted plane partitions, find the corresponding cyclically symmetric plane partition:

a
$$\begin{array}{cccc} 4 & 4 & 3 & 2 \\ & 3 & 2 & 1 \\ & & 1 & \end{array}.$$

b
$$\begin{array}{ccccc} 5 & 5 & 4 & 3 & 3 \\ & 4 & 3 & 2 & 1 \\ & & 2 & 1 & \end{array}.$$

c
$$\begin{array}{ccccccc} 7 & 7 & 7 & 4 & 3 & 3 & 3 \\ & 5 & 5 & 3 & 2 & 2 & \\ & & 4 & 2 & 1 & 1 & \end{array}.$$

1.3.2 For each of the following cyclically symmetric plane partitions, find the corresponding strict shifted plane partition.

a
$$\begin{array}{ccc} 3 & 3 & 2 \\ 3 & 3 & 2 \\ 2 & 2 & \end{array}.$$

b

$$\begin{array}{ccccc} 5 & 4 & 3 & 3 & 3 \\ 5 & 4 & 3 & 2 & \\ 5 & 3 & 3 & & \\ 2 & 2 & 1 & & \\ 1 & 1 & 1 & & \end{array} .$$

c

$$\begin{array}{ccccccc} 7 & 4 & 4 & 4 & 3 & 1 & 1 \\ 5 & 4 & 4 & 4 & & & \\ 5 & 4 & 4 & 4 & & & \\ 4 & 4 & 4 & 4 & & & \\ 1 & 1 & 1 & & & & \\ 1 & & & & & & \\ 1 & & & & & & \end{array} .$$

1.3.3 Without actually constructing the cyclically symmetric plane partition, find the number of cubes in the cyclically symmetric plane partitions that corresponds to each of the following strict shifted plane partitions:

a

$$\begin{array}{cccc} 4 & 3 & 3 & 1 \\ & 2 & 2 & \\ & & 1 & \end{array} .$$

b

$$\begin{array}{ccccccc} 7 & 7 & 6 & 5 & 4 & 4 & 3 \\ & 6 & 4 & 4 & 3 & 2 & 2 \\ & & 3 & 3 & 2 & & \\ & & & 2 & 1 & & \end{array} .$$

c

$$\begin{array}{cccccccccc} 10 & 9 & 9 & 8 & 8 & 8 & 7 & 5 & 4 & 4 \\ & 7 & 7 & 7 & 6 & 6 & 6 & 4 & & \\ & & 6 & 5 & 5 & 5 & 5 & 2 & & \\ & & & 4 & 4 & 4 & 3 & & & \\ & & & & 3 & 3 & 1 & & & \\ & & & & & 1 & & & & \end{array} .$$

1.3.4 Find a general rule to determine the total number of cubes in a cyclically symmetric plane partition that corresponds to a strict shifted plane partition with k rows and whose jth row is $a_{j,j}\ a_{j,j+1}\ \ldots\ a_{j,r_j}$ where $r_j = a_{j,j}$.

1.3.5 Prove that

$$\begin{bmatrix} a \\ j \end{bmatrix}_q = \begin{bmatrix} a-1 \\ j-1 \end{bmatrix}_q + q^j \begin{bmatrix} a-1 \\ j \end{bmatrix}_q .$$

This enables us to define the Gaussian polynomial recursively:

```
GP[a_,0,q_] = 1;
GP[a_,a_,q_] = 1;
GP[a_,j_,q_] := GP[a,j,q]
          = Expand[GP[a-1,j-1,q]+q^j GP[a-1,j,q]]
```

Find the Gaussian polynomials when $a = 4$, 5, and 6. What conjectures can you make about properties of Gaussian polynomials?

1.3.6 Prove that

$$\lim_{q\to 1} \begin{bmatrix} a \\ j \end{bmatrix}_q = \binom{a}{j}.$$

1.3.7 Use *Mathematica* to evaluate the determinant in equation (1.12) and verify that it agrees with the product representation of the generating function for cyclically symmetric plane partitions when r is less than or equal to 6.

1.3.8 Use *Mathematica* to evaluate the determinant in equation (1.14) and verify that it agrees with the product representation of the generating function for descending plane partitions when r is less than or equal to 6.

1.3.9 Prove that for all $r \geq 1$:

$$\prod_{1\leq i\leq j\leq r} \frac{r+i+j-1}{2i+j-1} = \prod_{j=0}^{r-1} \frac{(3j+1)!}{(r+j)!}.$$

1.3.10 Find the fourteen descending plane partitions with parts less than or equal to 4 and exactly one part of size 4.

1.3.11 Given a descending plane partition with parts less than or equal to r, if we remove the first row, we are left with a descending plane partition with parts less than or equal to $r-1$. We shall call this a **stripped descending plane partition**. It is a descending plane partition that could have been obtained by removal of the first row of a larger descending plane partition. Show that there are descending plane partitions that are not stripped. Describe necessary and conditions for a descending plane partition to be stripped.

1.3.12 Find the alternating sign matrix that corresponds to the square ice pattern

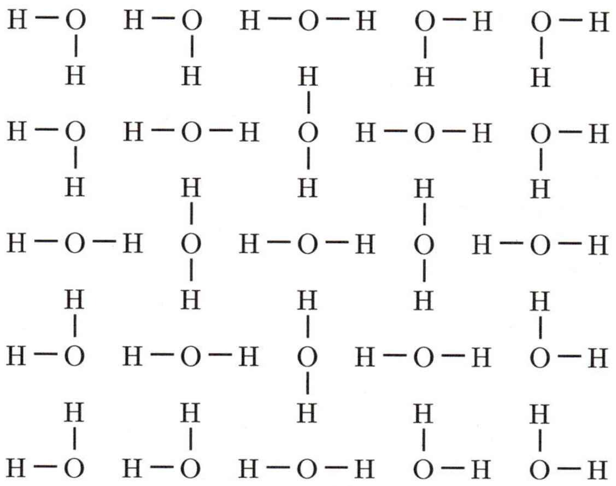

```
H—O   H—O   H—O—H   O—H   O—H
  |     |           |     |
  H     H     H     H     H
              |
H—O   H—O—H   O   H—O—H   O—H
  |           |           |
  H     H     H     H     H
        |           |
H—O—H   O   H—O—H   O   H—O—H
        |           |
  H     H     H     H     H
  |           |           |
H—O   H—O—H   O   H—O—H   O—H
              |
  H     H     H     H     H
  |     |           |     |
H—O   H—O   H—O—H   O—H   O—H
```

1.3.13 Find the configuration of square ice that corresponds to the alternating sign matrix

$$\begin{pmatrix} 0 & 0 & 1 & 0 & 0 & 0 \\ 0 & 1 & -1 & 0 & 1 & 0 \\ 1 & -1 & 0 & 1 & 0 & 0 \\ 0 & 1 & 0 & 0 & -1 & 1 \\ 0 & 0 & 1 & -1 & 1 & 0 \\ 0 & 0 & 0 & 1 & 0 & 0 \end{pmatrix}.$$

In 1875, the Johns Hopkins University, at Baltimore, was founded, and the Trustees sought the advice of the president, Daniel C. Gilman, in the selection of the professorial staff. He replied "Enlist a great mathematician and a distinguished Grecian; your problem will be solved. Such men can teach in a dwelling-house as well as in a palace. Part of the apparatus they will bring; part we will furnish. Other teachers will follow them." Joseph Henry also advised that liberal salaries should be paid and the best men in the world secured. He brought Sylvester's name prominently forward, and finally the latter was offered the post of Professor of Mathematics....

[Sylvester] found the conditions ideal. While not being overburdened with routine work, he was surrounded by able assistants and talented pupils only too eager to aid him in his profound original work or to catch inspiration from his lips. The mathematical staff was indeed very strong, including men of such capacity as Thomas Craig, W. E. Story, and Fabian Franklin. Sylvester's first class consisted of but one student, G. B. Halsted. This gentleman, since well known in science, had the most beneficial effect upon his master, for it was owing to his enthusiasm and persistence that Sylvester's attention was again called to the Modern Higher Algebra and the Theory of Invariants, and a fruitful crop of new discoveries was almost the immediate result. Others, including Franklin, Durfee, Ely, and Hammond in England joined in the investigations; a school of mathematics was founded; and the American renaissance in mathematics was an accomplished fact.

– P. A. MacMahon (1898)

2
Fundamental Structures

I sent to the *Comptes Rendus* two or three days ago my proof of the wonderful theorem (discovered by observation) [on] partitions of n into odd numbers and its partitions into unequal numbers. Franklin, Mrs. Franklin, Story, Hathaway, Ely, and Durfee† were all at work trying to find the proof – but I was fortunately beforehand with the theory and the only one in at the death.

– J. J. Sylvester to Arthur Cayley (1883)

Major MacMahon's investigations of plane partitions had grown out of his interest in ordinary partitions, representations of an integer as a sum of positive integers. These have a rich heritage going back at least to 1696 when Gottfried Wilhelm Leibniz wrote to Jean Bernoulli asking whether he knew how to enumerate the number of ways of writing a positive integer as a sum of one or more positive integers. The number 5 can be expressed in seven different ways as

$$5, \quad 4+1, \quad 3+2, \quad 3+1+1, \quad 2+2+1, \quad 2+1+1+1, \text{ or } 1+1+1+1+1.$$

These are the **partitions** of 5. The reason for Leibniz's interest is that every homogeneous symmetric polynomial of degree five in at least five variables is a linear combination of seven basic polynomials that are described by these seven partitions. He was interested in symmetric polynomials because they are the key to determining when the roots of a polynomial can be expressed as a simple function‡ of its coefficients.

Symmetric polynomials and partition theory are two of the basic themes that run through all aspects of the story of alternating sign matrices. MacMahon's work on plane partitions would eventually feed

† Except for William E. Story, who was an associate professor, these were Sylvester's graduate students at Johns Hopkins: Fabian Franklin, Christine Ladd Franklin, Arthur Safford Hathaway, George Stetson Ely, and William Pitt Durfee.

‡ That is to say, a formula in the coefficients involving only the four basic operations of arithmetic plus the taking of square roots.

into Alfred Young's investigations into the representation theory of the symmetric group, itself a direct successor of the study of symmetric polynomials. A **symmetric polynomial** in the variables $x_1, x_2, \ldots, x_n$ is a polynomial that is unchanged under any permutation of these variables. For example,

$$\begin{aligned} f(x, y, z) &= x^2y + x^2y^2 + x^2yz + x^2z + x^2z^2 + xy^2 + xy^2z \\ &\quad + xyz^2 + xz^2 + y^2z + y^2z^2 + yz^2 \end{aligned}$$

is a symmetric polynomial in x, y, and z:

$$f(x, y, z) = f(y, x, z) = f(y, z, x).$$

A **homogeneous symmetric polynomial** is a symmetric polynomial in which each monomial has the same degree. That is to say, the sum of the exponents is the same in each monomial. Our example is the sum of a homogeneous polynomial of degree three:

$$x^2y + x^2z + xy^2 + xz^2 + y^2z + +yz^2,$$

and a homogeneous polynomial of degree four:

$$x^2y^2 + x^2yz + x^2z^2 + xy^2z + xyz^2 + y^2z^2.$$

Once we know the first term of the homogeneous polynomial of degree 3, there are no surprises. Each of the other terms *must* be there if this polynomial is to be symmetric. For the homogeneous polynomial of degree four, everything is determined by the first two terms: x^2y^2 and x^2yz.

Given a weakly decreasing sequence of non-negative integers, $\lambda_1 \geq \lambda_2 \geq \cdots \geq \lambda_n \geq 0$, we define the **monomial symmetric function**, $m_{(\lambda_1,\lambda_2,\ldots,\lambda_n)}$, to be the smallest symmetric polynomial that contains the monomial $x_1^{\lambda_1} x_2^{\lambda_2} \cdots x_n^{\lambda_n}$. We can now write our example very succinctly as

$$f(x, y, z) = m_{(2,1)} + m_{(2,2)} + m_{(2,1,1)}.$$

If we have at least five variables, then there are seven monomial symmetric functions of degree five, one for each of the partitions of 5:

$$\begin{aligned} m_{(5)} &= x_1^5 + x_2^5 + \cdots, \\ m_{(4,1)} &= x_1^4 x_2 + \cdots, \\ m_{(3,2)} &= x_1^3 x_2^2 + \cdots, \\ m_{(3,1,1)} &= x_1^3 x_2 x_3 + \cdots, \end{aligned}$$

$$\begin{aligned} m_{(2,2,1)} &= x_1^2 x_2^2 x_3 + \cdots, \\ m_{(2,1,1,1)} &= x_1^2 x_2 x_3 x_4 + \cdots, \\ m_{(1,1,1,1,1)} &= x_1 x_2 x_3 x_4 x_5 + \cdots. \end{aligned}$$

The space of homogeneous symmetric polynomials in n variables has a linear basis given by the monomial symmetric functions indexed by the partitions of n.

2.1 Generating functions

So far as we know, Jean Bernoulli never gave Leibniz a good answer. It was Leonard Euler in the 1740s who discovered an efficient way of calculating $p(n)$, the number of partitions of n. His starting point is the generating function for partitions:

$$1 + \sum_{n=1}^{\infty} p(n)\, q^n = 1 + q + 2q^2 + 3q^3 + 5q^4 + 7q^5 + \cdots.$$

We define $p(0) = 1$; the empty partition is the one partition of 0.

This generating function can be written as an infinite product:

$$\sum_{n=0}^{\infty} p(n)\, q^n = \prod_{k=1}^{\infty} \frac{1}{1-q^k}, \qquad |q| < 1. \tag{2.1}$$

To see how this product arises, we begin with a different generating function. If we expand

$$(1+q)(1+q^2)(1+q^3)\cdots(1+q^m) = 1 + q + q^2 + 2q^3 + \cdots + q^{m(m+1)/2},$$

the coefficient of q^n will be the number of ways of expressing n as a sum of distinct integers from the set $\{1, 2, 3, \ldots, m\}$. If we want to allow up to three repetitions of any of these integers, then the generating function is given by

$$(1+q+q^2+q^3)(1+q^2+q^4+q^6)(1+q^3+q^6+q^9)\cdots(1+q^m+q^{2m}+q^{3m}).$$

If we want unlimited repetitions, each of the sums in parentheses becomes infinite:

$$(1+q+q^2+\cdots)(1+q^2+q^4+\cdots)(1+q^3+q^6+\cdots)\cdots(1+q^m+q^{2m}+\cdots).$$

We now restrict q to have absolute value less than 1 so that these series converge.† Since each infinite sum is a geometric series, we can simplify

† Strictly speaking, this is not necessary since we can treat them as formal series for which $1 + q + q^2 + q^3 + \cdots$ is defined to mean $1/(1-q)$.

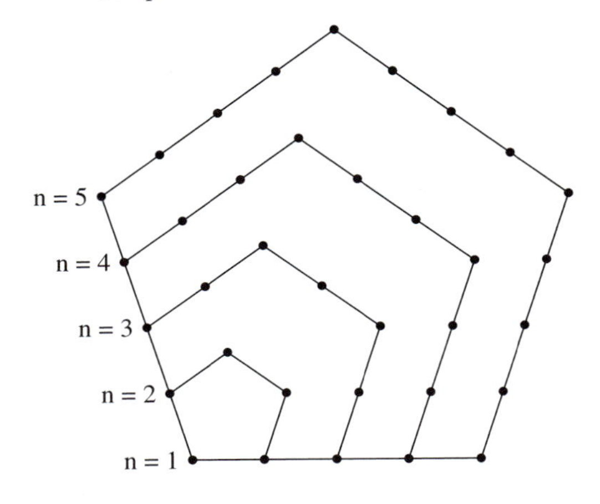

Figure 2.1. The pentagonal numbers.

this:

$$\frac{1}{1-q}\,\frac{1}{1-q^2}\,\frac{1}{1-q^3}\cdots\frac{1}{1-q^m}.$$

This is the generating function for the number of ways of writing an integer as a sum of integers from the set $\{1, 2, 3, \ldots, m\}$ with unlimited repetitions allowed. If we want to expand our set to include all possible integers, we need to take an infinite product.

At this point, Euler made a critical observation. If we expand the *reciprocal* of the generating function, we get a very interesting power series:

$$\prod_{k=1}^{\infty}(1-q^k) = 1 - q - q^2 + q^5 + q^7 - q^{12} - q^{15} + q^{22} + q^{26} - q^{35} - q^{40} + \cdots.$$

The coefficients appear to be always 0, 1, or -1, and most of them are 0. Furthermore, the powers of q with non-zero coefficients appear to come in pairs that each time are one further apart, and the first power in each pair falls into a sequence that Euler knew very well:

$$1, \quad 5, \quad 12, \quad 22, \quad 35, \quad \ldots\,.$$

These are the **pentagonal numbers**. The mth pentagonal number is the number of dots in a pentagonal arrangement with m dots on a side (see Fig. 2.1). The mth pentagonal number is $m(3m-1)/2$.

Theorem 2.1 (Euler's pentagonal number theorem)

$$\prod_{k=1}^{\infty}(1-q^k) = 1 + \sum_{m=1}^{\infty}(-1)^m \left[q^{m(3m-1)/2} + q^{m(3m+1)/2}\right]. \tag{2.2}$$

We shall postpone the proof of this theorem to Section 2.2. One of the reasons this theorem is important is that it gives us a recursive formula that enables us to calculate $p(n)$ efficiently.

Recursive formula for counting partitions

If we multiply the left side of equation (2.1) by the right side of equation (2.2), we get an infinite product times its reciprocal,

$$\left(\sum_{n=0}^{\infty} p(n)\, q^n\right)\left(1 + \sum_{m=1}^{\infty}(-1)^m \left[q^{m(3m-1)/2} + q^{m(3m+1)/2}\right]\right) = 1. \tag{2.3}$$

When we multiply two power series, we get

$$\sum_{n=0}^{\infty} a_n\, q^n \sum_{m=0}^{\infty} b_m\, q^m = \sum_{t=0}^{\infty} c_t\, q^t,$$

where $c_t = a_t b_0 + a_{t-1} b_1 + a_{t-2} b_2 + \cdots + a_0 b_t$. The coefficient of q^t on the left-hand side of equation (2.3) is

$$p(t) + \sum_{m \geq 1}(-1)^m \left[p(t - m(3m-1)/2) + p(t - m(3m+1)/2)\right].$$

We define the number of partitions of a negative integer to be zero and take the sum over all values of m for which the summand is not zero. The coefficient of q^t on the right-hand side of equation (2.3) is zero for all $t > 0$:

$$p(t) = \sum_{m \geq 1}(-1)^{m-1} \left[p(t - m(3m-1)/2) + p(t - m(3m+1)/2)\right], \ t > 0. \tag{2.4}$$

We know $p(n)$ for $0 \leq n \leq 5$. To find $p(6)$ through $p(12)$, we use this recursive formula:

$$\begin{aligned} p(6) &= p(5) + p(4) - p(1) \\ &= 7 + 5 - 1 = 11, \\ p(7) &= p(6) + p(5) - p(2) - p(0) \\ &= 11 + 7 - 2 - 1 = 15, \end{aligned}$$

$$\begin{aligned}
p(8) &= p(7)+p(6)-p(3)-p(1) \\
&= 15+11-3-1=22, \\
p(9) &= p(8)+p(7)-p(4)-p(2) \\
&= 22+15-5-2=30, \\
p(10) &= p(9)+p(8)-p(5)-p(3) \\
&= 30+22-7-3=42, \\
p(11) &= p(10)+p(9)-p(6)-p(4) \\
&= 42+30-11-5=56, \\
p(12) &= p(11)+p(10)-p(7)-p(5)+p(0) \\
&= 56+42-15-7+1=77.
\end{aligned}$$

Another use for generating functions

Euler realized that generating functions could do more than give recursive formulæ; they could also be used to prove that two sets had the same cardinality. Consider the set of partitions of n into odd parts. For $n = 5$, there are three partitions:

$$5, \quad 3+1+1, \quad \text{and} \quad 1+1+1+1+1.$$

The generating function for $p_{\mathcal{O}}(n)$, the number of ways of writing n as a sum of positive odd integers, is

$$1+\sum_{n=1}^{\infty} p_{\mathcal{O}}(n)\, q^n = \frac{1}{1-q}\,\frac{1}{1-q^3}\,\frac{1}{1-q^5}\cdots.$$

We now manipulate this generating function. The following steps can be justified by taking finite products and then passing to the limit:

$$\begin{aligned}
\frac{1}{1-q}\,\frac{1}{1-q^3}\,\frac{1}{1-q^5}\cdots &= \frac{1}{1-q}\frac{1-q^2}{1-q^2}\frac{1}{1-q^3}\frac{1-q^4}{1-q^4}\frac{1}{1-q^5}\frac{1-q^6}{1-q^6}\cdots \\
&= \frac{1-q^2}{1-q}\,\frac{1-q^4}{1-q^2}\,\frac{1-q^6}{1-q^3}\,\frac{1-q^8}{1-q^4}\cdots \\
&= (1+q)(1+q^2)(1+q^3)(1+q^4)\cdots. \qquad (2.5)
\end{aligned}$$

We have already seen that this last product is the generating function for the number of partitions of n into distinct parts, $p_{\mathcal{D}}(n)$:

$$(1+q)(1+q^2)(1+q^3)(1+q^4)\cdots = 1+\sum_{n=1}^{\infty} p_{\mathcal{D}}(n)\, q^n.$$

It follows that

$$1+\sum_{n=1}^{\infty} p_{\mathcal{O}}(n)\, q^n = 1+\sum_{n=1}^{\infty} p_{\mathcal{D}}(n)\, q^n,$$

and therefore we have proven the following theorem.

Theorem 2.2 *The number of partitions of n into odd parts is equal to the number of partitions of n into distinct parts:*

$$p_{\mathcal{O}}(n) = p_{\mathcal{D}}(n). \tag{2.6}$$

We have seen that $p_{\mathcal{O}}(5) = 3$. There are also three partitions of 5 into distinct parts:

$$5, \quad 4+1, \quad \text{and} \quad 3+2.$$

Extensions and generalizations

This is our first complete proof. It provides us with an opportunity to reflect on what has been accomplished. I hope that you now believe that partitions of n into odd parts are equinumerous with partitions of n into distinct parts, but I probably could have convinced you with overwhelming numerical evidence. What is more significant is that you have seen a technique that is easily generalized. What happens if we take a finite product,

$$\frac{1}{1-q}\,\frac{1}{1-q^3}\cdots\frac{1}{1-q^{2m-1}} = \frac{(1+q)(1+q^2)\cdots(1+q^m)}{(1-q^{m+1})(1-q^{m+2})\cdots(1-q^{2m})}?$$

Corollary 2.3 *The number of partitions of n into odd parts less than or equal to $2m-1$ is equal to the number of partitions of n into parts less than or equal to $2m$ for which the parts must be distinct if they are less than or equal to m.*

What happens if instead of omitting even parts we omit multiples of 3,

$$\begin{aligned}
&\frac{1}{1-q}\,\frac{1}{1-q^2}\,\frac{1}{1-q^4}\,\frac{1}{1-q^5}\,\frac{1}{1-q^7}\cdots \\
&\qquad = \frac{1-q^3}{1-q}\,\frac{1-q^6}{1-q^2}\,\frac{1-q^9}{1-q^3}\,\frac{1-q^{12}}{1-q^4}\cdots \\
&\qquad = (1+q+q^2)(1+q^2+q^4)(1+q^3+q^6)(1+q^4+q^8)\cdots?
\end{aligned}$$

Corollary 2.4 *The number of partitions of* n *into parts not divisible by* 3 *is equal to the number of partitions of* n *in which each integer appears at most twice.*

What happens if we omit multiples of d?

In the next section, we shall see a very different proof of Theorem 2.2 that has its own extensions and generalizations.

Exercises

2.1.1 Prove that if $f(x, y, z) = f(y, x, z) = f(y, z, x)$, then f is symmetric in x, y, and z.

2.1.2 If we have three variables, how many distinct monomial symmetric functions of degree five are there? How many distinct monomial symmetric functions are there of degree six in three variables?

2.1.3 Find and prove a formula for the number of distinct monomial symmetric functions of degree n in two variables.

2.1.4 Find and prove a formula for the number of distinct monomial symmetric functions of degree n in three variables.

2.1.5 Let $a_k(n)$ be the number of ways of rolling k distinct dice to get a total of n. For example, $a_2(2) = 1$, $a_2(3) = 2$, $a_2(4) = 3, \ldots, a_2(12) = 1$. Explain why the generating function for $a_k(n)$ is equal to $(x - x^7)^k/(1 - x)^k$.

2.1.6 If we use dice in which one side has a 1, two sides have 2, and three sides have 3, find the generating function for the number of ways of rolling n using k dice.

2.1.7 Write down all the partitions of 6. Write down all the partitions of 10 into distinct parts. Write down all the partitions of 10 into odd parts.

2.1.8 Write a program that will list all partitions of n.

2.1.9 How many dots do you have to add to a pentagon with $m-1$ dots on a side to get a pentagon with m dots on a side? Prove by induction that the mth pentagonal number is $m(3m-1)/2$.

2.1.10 Let j and k be two positive integers. Show that

$$\frac{3j^2 \pm j}{2} = \frac{3k^2 \pm k}{2}$$

if and only if $j = k$. You need to verify this when you have a plus sign on both sides or a minus sign on both sides. Show that if you have a plus sign on one side and a minus sign on the other, then you cannot have equality.

2.1.11 Write a program that uses the recursion in equation (2.4) to find the number of partitions of n for all $n \leq 100$. Use this table to conjecture a *sufficient* condition for $p(n)$ to be divisible by 5.

2.1.12 Let $L(\pi)$ be the number of parts in π, called the **length** of π. Let $|\pi|$ be the sum of the parts in π. Let $\mathcal{P}_\mathcal{D}$ be the set of partitions into distinct parts. Prove that

$$(1-x)(1-x^2)(1-x^3)\cdots = \sum_{\pi \in \mathcal{P}_\mathcal{D}} (-1)^{L(\pi)} q^{|\pi|}.$$

2.1.13 Let $\mathcal{P}$ be the set of all partitions. Prove that

$$\frac{1}{(1-tx)(1-tx^2)(1-tx^3)\cdots} = \sum_{\pi \in \mathcal{P}} t^{L(\pi)} q^{|\pi|}.$$

2.1.14 The following *Mathematica* command, `ModProd[l,m,n]`, produces the terms up to q^n of the expansion of $\prod(1-q^j)$ where j is restricted to lie in one of the residue classes in the set l (of positive integers) modulo m:

```
ModProd[l_,m_,n_] := Series[ Product[
                     Product[1-q^(l[[i]]+m*t),
                     {i,Length[l]}],{t,0,n}],{q,0,n}]
```

Evaluate `ModProd[{1,3,4},4,n]` for various values of n and conjecture a summation formula for $\prod(1-q^i)$, $i \geq 1$, $i \not\equiv 2 \pmod 4$.

2.1.15 Find other sets of residue classes for other moduli for which you can conjecture summation formulæ.

2.1.16 Explain why the coefficient of q^n in $\prod_{j\geq 1}(1-q^j)$ is equal to the number of ways writing n as a sum of an even number of distinct positive integers minus the number of ways of writing n as a sum of an odd number of distinct positive integers.

2.1.17 Prove that for any integer $d \geq 2$, the number of partitions of n into parts not divisible by d is equal to the number of partitions of n in which each integer appears less than d times.

2.1.18 Prove Theorem 2.1 by proving that if $n = 6j^2 \pm j$ for some integer j, then there is one more representation of n as a sum of an even number of distinct positive integers. If $n = (2j+1)(3j \pm 1)$ for some integer j, then there is one more representation of n as a sum of an odd number of distinct positive integers. And in all other cases, there are the same number of representations.

2.2 Partitions

James Joseph Sylvester (1814–1897) was one of the most influential mathematicians of the nineteenth century. Although he attended Cambridge, he was not permitted to receive his degree because he was Jewish and therefore unwilling to embrace the thirty-nine articles of the Church of England. For the same reason, he was unable to teach in any of the chartered universities, but gained employment at University College in the University of London which had just been founded, largely to provide education for English Jews, Catholics, Unitarians, and other dissenters. He received his MA from Trinity College, Dublin, in 1841, then went to the University of Virginia for a few turbulent months. Returning to England, he worked as an actuary, passed the bar in 1850, and in 1855 began teaching at the Royal Military Academy in Woolwich. Forced into retirement at the age of fifty-five, he began a fresh career at Johns Hopkins University. In 1883, he once again returned to England to become Savilian Professor of Mathematics at Oxford.

Sylvester was responsible for much of matrix theory – including the coining of the mathematical term "matrix" in 1850 – as well as for much of the invariant theory that would eventually lead to results such as those on Schur functions which we shall see in Chapter 4. But our interest in Sylvester rests on his work in partitions, especially what he accomplished in his unconventional memoir "A Constructive Theory of Partitions, Arranged in Three Acts, an Interact, and an Exodion" (1882). We shall examine a bijective proof of Theorem 2.2 that comes from this paper and which will lead us into a proof of Jacobi's identity. This paper is pure Sylvester in its exuberance and indulgence in flowery oratory. It is the fruit of a seminar he taught at the newly founded Johns Hopkins University, and contributions from his students weave in and out of the exposition.

Figure 2.2. James Joseph Sylvester (From the Ferdinand Hamburger, Jr. Archives of the Johns Hopkins University).

There are many natural bijections between partitions into odd parts and partitions into distinct parts. In exercise 2.2.6, we shall see a different bijection that was published by James Whitbread Lee Glaisher, a Cambridge mathematician, in 1883. Glaisher is remembered for his work on representations of an integer as a sum of squares. Among his popular accomplishments was the solution of the n queens problem: In how many ways can n queens be placed on an $n \times n$ chess board so that no two attack each other?

Ferrers graphs

Norman MacLeod Ferrers (1829–1903) was a Fellow and Tutor of Gonville and Caius College, Cambridge. He became Master of the College in 1880 and Vice-Chancellor of the University in 1884. Better known as a teacher and administrator than as a research mathematician, he nevertheless did important work on spherical harmonics. In the 1850s, he wrote to Sylvester with a simple proof of an observation made by Euler, that the number of partitions of n into exactly m parts is equal to the number of partitions of n whose largest part is m. For example, there are four partitions of 7 into three parts:

$$5+1+1, \quad 4+2+1, \quad 3+2+2, \quad \text{and} \quad 3+3+1,$$

and there are four partitions of 7 whose largest part is 3:

$$3+1+1+1, \quad 3+2+1+1, \quad 3+3+1, \quad \text{and} \quad 3+2+2.$$

If we represent each part as a row of dots, the first set of partitions is

• • • • • • •	• • • • • • •	• • • • • • •	• • • • • • •

The second set of partitions consists of

• • • • • • •	• • • • • • •	• • • • • • •	• • • • • • •

Each partition in the second set is the reflection across the main diagonal (running northwest to southeast) of the partition that lies directly above it. In general, if we take a partition with exactly m parts and reflect it about the main diagonal, we get a unique partition whose largest part is m, and vice versa. This establishes a one-to-one correspondence between these two sets of partitions.

Such diagrams are called **Ferrers graphs**, and the act of reflecting a Ferrers graph across the main diagonal is called **conjugation**.

What we have just seen is a proof by example, one of the most primitive and unreliable and yet appealing methods of proof. Any proof by example relies on the acceptance that the example is truly generic. In this case, we must be convinced that for any m and for any partition into exactly m parts, this process that we have called conjugation will always produce a well-defined partition with largest part m, and that conjugation is uniquely reversible. Our next proof is another proof by example, but one of greater subtlety.

Sylvester's bijection

To get Sylvester's bijection between partitions of n into odd parts and partitions of n into distinct parts, we start with a partition into odd parts and modify its Ferrers graph so that the central dots in each part all lie in the same column. For example, we represent the partition $7+7+5+5+3+1+1+1$ of 30 as

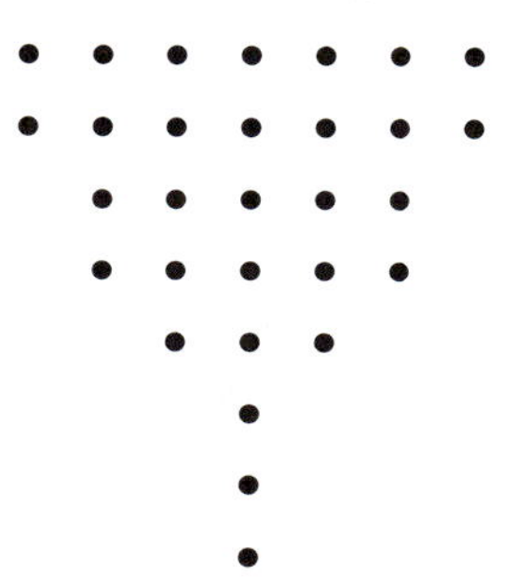

We now draw an angled line up the center column to the top row and then over to the right. The dots on this line represent the first part in our partition into distinct parts. There are eleven of them. For the second part, we draw a line up the column to the left of center to the first row and then over to the left. This gives us seven as our second part. For the third part, we go up the column to the right of the center as far as the second row (so that we do not go through any dot twice) and then over to the right. Our third part is six. We continue alternating sides of the center column, taking the closest column to the center which has not yet been counted, drawing a line up to the highest row which is not yet completely taken, and then moving off to the left or right, depending on which side of the central column we are on.

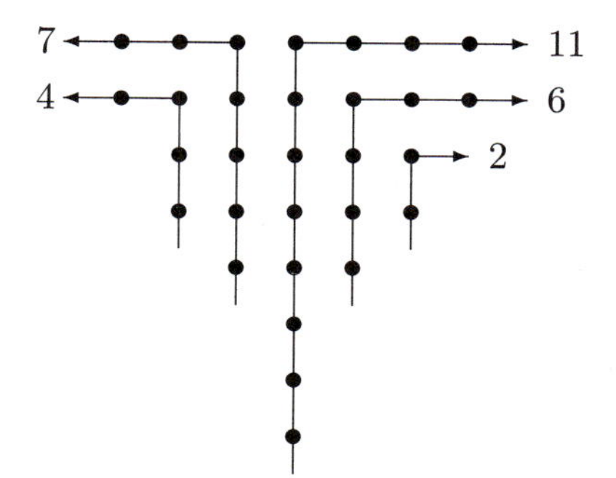

This gives us the partition $11 + 7 + 6 + 4 + 2$ in which the parts are distinct. To see that this really is a bijection, we need to explain how to reverse this process. Given a partition into t distinct parts,

1. if t is odd, then the smallest part corresponds to a column of dots on the right,
2. if t is even, then the smallest part corresponds to a row of dots on the left,

3. each time we represent a part by an angle of dots on the right, the next larger part is represented by an angle on the left with a column of dots that is one longer than that of the column that has just been placed,
4. each time we represent a part by an angle of dots on the left, the next larger part is represented by an angle on the right with a row of dots that is one longer than that of the row that has just been placed.

This establishes the bijection.

The field of combinatorics is sufficiently new and varied that there are many instances where proof by example is still the best proof available. While more rigorous proofs can be given, they begin to obscure the simple patterns and relationships that the proof is intended to illuminate. As Sylvester gave his bijection and as I have recreated it here, the proof is sketchy. You should be skeptical. The exercises at the end of this section will give you the opportunity to play with this bijection and convince yourself that it is, indeed, generic.

The triple product identity

Carl Gustav Jacob Jacobi (1804–1851) will make several contributions to our story. Like Sylvester, he was Jewish. He went to the University of Berlin where he graduated in 1824, but in order to obtain an academic position he chose to convert to Christianity. Like Sylvester, Jacobi was instrumental in the development of matrix theory and the invariant analysis that accompanies it. Also as with Sylvester, we shall begin by looking at a different but not unrelated subject on which he worked: in this case, elliptic functions.

In the eighteenth century, many mathematicians had wrestled with integrals of the form

$$u(x) = \int_0^x \frac{p(t)}{\sqrt{q(t)}}\, dt$$

where p and q are polynomials and q has degree three or four. The arc length of a portion of an ellipse gives rise to one of these integrals, hence the name elliptic integral. Adrien-Marie Legendre did a great deal of preparatory work on their evaluation, but in the 1820s it was two young mathematicians, Gustav Jacobi and Niels Henrik Abel, who discovered the critical insight that one must treat x as a complex variable and then look at the inverse function, x as a function of u. In 1829, Jacobi traveled

Figure 2.3. Carl Gustav Jacob Jacobi.

to Paris to meet Legendre and the other Parisian mathematicians, stopping off in Göttingen to talk with Gauss. In the same year, he published his great treatise on elliptic functions, *Fundamenta nova theoriæ functionum ellipticarum.* Euler's pentagonal number theorem, Theorem 2.1, is a special case of an identity that Jacobi presented, proved, and used in his thesis, the triple product identity.

We begin with an infinite product,

$$\begin{aligned} f(x) &= \prod_{k=1}^{\infty}(1+xq^k)(1+x^{-1}q^{k-1}) \\ &= (1+xq)(1+xq^2)(1+xq^3)\cdots \\ &\quad \times (1+x^{-1})(1+x^{-1}q)(1+x^{-1}q^2)\cdots. \end{aligned}$$

This infinite product converges when $|q| < 1$. It can be expanded as a **Laurent series** in x, which means as a power series in which both positive and negative powers of x appear. The coefficient of each power of x will be a function of q:

$$f(x) = \sum_{n=-\infty}^{\infty} a_n x^n, \qquad a_n = a_n(q). \tag{2.7}$$

The symmetry of this function tells us a great deal about these coefficients. In particular, we have that

$$\begin{aligned} f(xq) &= (1+xq^2)(1+xq^3)(1+xq^4)\cdots \\ &\quad \times (1+x^{-1}q^{-1})(1+x^{-1})(1+x^{-1}q)\cdots \\ &= \frac{1+x^{-1}q^{-1}}{1+xq} \times (1+xq)(1+xq^2)(1+xq^3)\cdots \\ &\quad \times (1+x^{-1})(1+x^{-1}q)(1+x^{-1}q^2)\cdots \\ &= x^{-1}q^{-1}\, f(x). \end{aligned}$$

From this equation, it follows that

$$\sum_{n=-\infty}^{\infty} a_n x^n q^n = \sum_{n=-\infty}^{\infty} a_n x^{n-1} q^{-1}.$$

If we compare the coefficients of x^n on each side, we see that

$$a_n q^{n+1} = a_{n+1}. \tag{2.8}$$

If we know a_0, then we can find a_n when n is positive:

$$a_1 = qa_0, \quad a_2 = q^2 a_1 = q^3 a_0, \quad a_3 = q^3 a_2 = q^6 a_0, \quad \ldots,$$
$$a_n = q^n a_{n-1} = q^{1+2+3+\cdots+n} a_0 = q^{n(n+1)/2} a_0.$$

We can also find a_n when n is negative by rewriting equation (2.8) as $a_{n-1} = q^{-n} a_n$:

$$a_{-1} = q^0 a_0 = a_0, \quad a_{-2} = q^1 a_{-1} = qa_0, \quad a_{-3} = q^2 a_{-2} = q^3 a_0,$$
$$\ldots, \quad a_{-n} = q^{n-1} a_{-n+1} = q^{1+2+\cdots+(n-1)} a_0 = q^{(-n)(-n+1)/2} a_0.$$

Putting this all together, we find that

$$f(x) = a_0 \sum_{n=-\infty}^{\infty} q^{n(n+1)/2} x^n. \tag{2.9}$$

Finding a_0

All that is left is to find a_0, which we know is a function of q. It is the coefficient of x^0 in the expansion of the infinite product

$$(1+xq)(1+xq^2)(1+xq^3)\cdots(1+x^{-1})(1+x^{-1}q)(1+x^{-1}q^2)\cdots.$$

In other words, a_0 consists of those terms in which the powers of x cancel. It will be a power series in q in which the coefficient of q^m is the number of ways of getting a q^m by taking distinct terms, xq^i, from the first infinite product and an equal number of distinct terms, $x^{-1}q^j$, from the second infinite product so that the sum of the powers of q equals m.

As an example, the coefficient of q^3 in a_0 is 3:

$$xq^3 \times x^{-1} = q^3, \quad xq^2 \times x^{-1}q = q^3, \quad xq \times x^{-1}q^2 = q^3.$$

The coefficient of q^4 is 5. There are four ways of getting q^4 by taking one term from the first infinite product and one term from the second infinite product. In addition, we can take the first two terms from each of the infinite products:

$$xq \times xq^2 \times x^{-1} \times x^{-1}q = q^4.$$

The following lemma summarizes this technique for calculating the coefficients of q in a_0.

Lemma 2.5 *The coefficient $a_0(q)$ is a power series in q:*

$$a_0 = \sum_{m=0}^{\infty} b_m q^m,$$

where b_m is the number of ways of representing m as a sum of distinct elements from the set $\{1, 2, 3, \ldots\}$, plus an equal number of distinct elements taken from the set $\{0, 1, 2, \ldots\}$.

If we work out the first few coefficients, we see that

$$a_0 = 1 + q + 2q^2 + 3q^3 + 5q^4 + 7q^5 + \cdots.$$

By now, this series should be familiar, at least suggesting the following lemma.

Lemma 2.6 *The coefficient of q^m in $a_0(q)$ is the number of partitions of m.*

Proof: We need to show that the number of partitions of m is always the same as the number of ways of representing m as a sum of distinct positive integers plus an equal number of distinct non-negative integers (in other words, we allow up to one 0). The Ferrers graph of the partition shows us how to do this. We draw a diagonal line just below the dots along the main diagonal. The lengths of the rows to the right of this line give us the positive integers, the lengths of the columns below this line give us the non-negative integers. As an example, the partition 6+5+5+4+2+1 of 23 breaks down as follows:

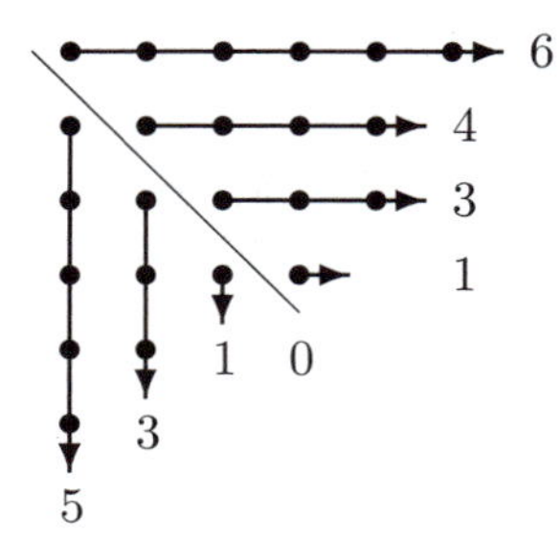

We represent 23 as $6 + 4 + 3 + 1$ from the first set plus $5 + 3 + 1 + 0$ from the second set. It should be clear that each such a pair of sums corresponds to a unique partition.

Q.E.D.

Since $a_0(q)$ is the generating function for partitions, we know that

$$a_0 = \prod_{j=1}^{\infty} \frac{1}{1 - q^j}.$$

We have proven the Jacobi triple product identity:

Theorem 2.7 (Jacobi triple product identity) *For $|q| < 1$ and any x, we have that*

$$\prod_{i=1}^{\infty} (1 + xq^i)(1 + x^{-1}q^{i-1}) = \prod_{j=1}^{\infty} \frac{1}{1 - q^j} \sum_{n=-\infty}^{\infty} q^{n(n+1)/2} x^n. \tag{2.10}$$

The name derives from the fact that this identity is usually written with all of the infinite products on the same side:

$$\prod_{i=1}^{\infty} (1 + xq^i)(1 + x^{-1}q^{i-1})(1 - q^i) = \sum_{n=-\infty}^{\infty} q^{n(n+1)/2} x^n. \tag{2.11}$$

This proof of the Jacobi triple product identity illustrates two points. The first is the power of symmetry arguments as shown by the translation symmetry between $f(x)$ and $f(xq)$ that leads to equation (2.9). As you will see in the exercises, this approach is easily extended to obtain series expansions for other products. Exploitation of symmetry will be a recurrent theme throughout this book, yielding many of our simplest and most insightful proofs.

The second point is that good proofs are often hybrids, in this case a marriage of the purely algebraic argument that leads to equation (2.9) and the combinatorial argument that proves Lemma 2.6. There are other ways to evaluate $a_0(q)$, but none as appealing. To those familiar with symmetry arguments and with partitions, each piece of this proof is immediate and transparent and suggests generalizations and connections. Most of the significant proofs in the remainder of this book will be hybrids.

Proof of the pentagonal number theorem

To prove Euler's pentagonal number theorem, equation (2.2), we start with equation (2.11), replace each q by q^3,

$$\prod_{i=1}^{\infty}(1+xq^{3i})(1+x^{-1}q^{3i-3})(1-q^{3i}) = \sum_{n=-\infty}^{\infty} q^{(3n^2+3n)/2}x^n,$$

and then set $x = -q^{-1}$:

$$\prod_{i=1}^{\infty}(1-q^{3i-1})(1-q^{3i-2})(1-q^{3i}) = \sum_{n=-\infty}^{\infty} (-1)^n q^{(3n^2+n)/2}. \qquad (2.12)$$

The product on the left can be rewritten as $\prod_{i=1}^{\infty}(1-q^i)$.

Q.E.D.

The simplest proofs of the pentagonal number theorem all proceed by first proving Jacobi's triple product identity, which is easier to prove because it has more structure.

Exercises

2.2.1 A partition is **self-conjugate** if it is equal to its conjugate partition. Use the Ferrers graph to prove that the number of self-conjugate partitions of n is the same as the number of partitions of n into distinct odd parts.

2.2.2 Let λ be a partition with parts $\lambda_1 \geq \lambda_2 \geq \cdots$. Define $\delta(\lambda)$ to be the largest j for which $\lambda_j \geq j$ and $\sigma(\lambda)$ to be the largest j for which $\lambda_j = \lambda_1 - j + 1$. Calculate the values of these functions for the partitions $6+5+3+2+1$, $7+6+3$, and $(2k-1)+(2k-2)+\cdots+(k+1)+k$. Describe these functions in terms of the Ferrers graph.

2.2.3 Find the partitions of 50 into odd parts that corresponds under

Sylvester's bijection to the partition of 50 into distinct parts given by

$$50 = 12 + 10 + 9 + 8 + 6 + 5.$$

2.2.4 Show that under Sylvester's bijection, the number of distinct odd integers which appear in the partition into odd parts is always equal to the number of runs in the partition into distinct parts, where a run is a maximal sequence of parts that are consecutive integers. For example, $12+10+9+8+6+5$ has three runs: $\{12\}$, $\{10, 9, 8\}$, $\{6, 5\}$.

2.2.5 Show that under Sylvester's bijection, the total number of odd parts is equal to the alternating sum of the corresponding distinct parts: $\lambda_1 - \lambda_2 + \lambda_3 - \cdots$, where $\lambda_1 > \lambda_2 > \lambda_3 > \cdots$ are the distinct parts. For example, $12 - 10 + 9 - 8 + 6 - 5 = 4$. The corresponding partition into odd parts must be a sum of four integers.

2.2.6 Glaisher's bijective proof that the number of partitions of n into odd parts is equal to the number of partitions of n into distinct parts proceeds by counting the number of times each odd part appears, representing that count as a sum of powers of 2, and then multiplying each of those powers of 2 by the odd number being counted. For example,

$$\begin{aligned}
&7+7+7+5+5+5+5+5+3+3+3+3+3+3+1+1 \\
&\quad = (2+1)7 + (4+1)5 + (4+2)3 + (2)1 \\
&\quad = 14 + 7 + 20 + 5 + 12 + 6 + 2.
\end{aligned}$$

Show that Glaisher's correspondence is uniquely reversible, establishing a one-to-one correspondence between the two sets of partitions. Prove that this correspondence is *not* the same as Sylvester's correspondence.

2.2.7 Generalize Glaisher's correspondence† to prove that the number of partitions of n in which each part appears less than d times is equal to the number of partitions of n in which there are no multiples of d.

2.2.8 Let $a(m, n)$ be the number of representations of m as the sum of integers in two partitions into distinct parts for which the first partition must have exactly n more elements than the second,

† This shows that Glaisher's proof is, in some sense, closer to the generating function proof that we saw in Section 2.1. A fruitful line of inquiry has been to seek the generating function proof that corresponds to Sylvester's bijection.

and the second is allowed to have 0s. If n is negative, then the second partition has exactly $-n$ more elements than the first. The coefficient $a_n = a_n(q)$ defined in equation (2.7) on page 47 is also given by

$$a_n(q) = \sum_{m=0}^{\infty} a(m,n)q^m.$$

Find a bijection between the partitions counted by $a(m,n)$ and those counted by $a(m-n,-n)$. Use the equal cardinality of these two sets to prove that $a_n(q) = q^n a_{-n}(q)$.

2.2.9 Continuing exercise 2.2.8, find a bijection between the partitions counted by $a(m,-n)$ and those counted by $a(m,n-1)$. Use this to prove that $a_{-n}(q) = a_{n-1}(q)$. Combine this result with the previous exercise to prove that

$$a_n = a_0 q^{n(n+1)/2}.$$

2.2.10 Prove that

$$\prod_{j=1}^{\infty}(1-q^{4j-3})(1-q^{4j-1})(1-q^{4j}) = \sum_{n=-\infty}^{\infty}(-1)^n q^{2n^2+n}.$$

2.2.11 Use the Jacobi triple product identity to prove that for integers a and m, $1 \leq a < m/2$:

$$\begin{aligned}\prod_{j=1}^{\infty}&(1-q^{mj-m+a})(1-q^{mj-a})(1-q^{mj}) \\ &= \sum_{n=-\infty}^{\infty}(-1)^n q^{(mn^2+(m-2a)n)/2}.\end{aligned}$$

2.2.12 Consider the function

$$\begin{aligned} g(x) &= \prod_{k=1}^{\infty}(1-xq^k)(1-x^{-1}q^{k-1}) \\ &\qquad \times (1-x^2q^{2k-1})(1-x^{-2}q^{2k-1}) \\ &= \sum_{n=-\infty}^{\infty} b_n\, x^n. \end{aligned}$$

By considering $g(x/q)$ and $g(x^{-1})$ in terms of $g(x)$, show that

$$b_n = q^{n-1}b_{n-3}, \quad \text{and} \quad b_{-n} = -b_{n-1}.$$

2.2.13 Continuing exercise 2.2.12, prove that $b_1 = 0$ and $b_{-1} = -b_0$, and therefore

$$g(x) = b_0(q) \sum_{n=-\infty}^{\infty} q^{(3n^2+n)/2}(x^{3n} - x^{-3n-1}).$$

2.2.14 Continuing exercise 2.2.12, set $x = -q$ and use the Jacobi triple product identity to show that

$$\begin{aligned}\prod_{k=1}^{\infty}&(1+q^{k+1})(1+q^{k-2})(1-q^{2k+1})(1-q^{2k-3}) \\ &= b_0(q)\left[\prod_{k=1}^{\infty}(1-q^{3k-5})(1-q^{3k+2})(1-q^{3k})\right. \\ &\qquad \left. + q^{-1}\prod_{k=1}^{\infty}(1-q^{3k-4})(1-q^{3k+1})(1-q^{3k})\right]. \qquad (2.13)\end{aligned}$$

2.2.15 Continuing exercise 2.2.12, use the fact that $\prod(1+q^k) = \prod 1/(1-q^{2k-1})$ to show that equation (2.13) is implies that $b_0(q) = \prod_{k=1}^{\infty}(1-q^k)^{-1}$, and therefore

$$\begin{aligned}\prod_{k=1}^{\infty}&(1-xq^k)(1-x^{-1}q^{k-1})(1-x^2q^{2k-1})(1-x^{-2}q^{2k-1})(1-q^k) \\ &= \sum_{n=-\infty}^{\infty} q^{(3n^2+n)/2}(x^{3n} - x^{-3n-1}). \qquad (2.14)\end{aligned}$$

This is called the **quintuple product identity**.

2.2.16 Use the quintuple product identity to find the series expansion of $\prod(1-q^k)$ where the product is taken over $k \geq 1$ for which $k \equiv 0, m, \pm a, m \pm a, \text{or } m \pm 2a \pmod{2m}$, provided $1 \leq a < m/2$ and $3a \neq m$.

2.2.17 The following identity is due to Lasse Winquist (1969)†:

$$\begin{aligned}\prod_{k=1}^{\infty}&(1-q^k)^2(1-xq^{k-1})(1-x^{-1}q^k)(1-yq^{k-1})(1-y^{-1}q^k) \\ &\times (1-xy^{-1}q^{k-1})(1-xyq^{k-1})(1-x^{-1}yq^k)(1-x^{-1}y^{-1}q^k)\end{aligned}$$

† Following Winquist's discovery, several mathematicians including Carlitz and Subbarao (1972), Dyson (1972), Macdonald (1972), and Stanton (1986, 1989) saw that this approach could be generalized. Infinite families of such product identities are now known.

$$= \sum_{i=-\infty}^{\infty} \sum_{j=-\infty}^{\infty} (-1)^{i+j} \left[x^{3i}(y^{3i} - y^{1-3i}) - y^{3i-1}(x^{3j+1} - x^{2-3j})\right]$$
$$\times\, q^{3i(i-1)/2+j(3j-1)/2}. \tag{2.15}$$

To prove it, begin by defining

$$\begin{aligned} t(x,y) &= \prod_{k=1}^{\infty}(1-xq^{k-1})(1-x^{-1}q^k)(1-yq^{k-1})(1-y^{-1}q^k) \\ &\qquad \times (1-xy^{-1}q^{k-1})(1-xyq^{k-1}) \\ &\qquad \times (1-x^{-1}yq^k)(1-x^{-1}y^{-1}q^k) \\ &= \sum_{m=-\infty}^{\infty} \sum_{n=-\infty}^{\infty} c_{m,n}(q)\, x^m y^n. \end{aligned}$$

Use the symmetries of $t(x,y)$ to prove that

$$\begin{aligned} c_{m,n} &= -c_{3-m,n}, \\ c_{m,n} &= -c_{n+1,m-1}, \\ c_{m,n} &= -q^{m-3} c_{m-3,n}, \\ c_{m,n} &= -q^{n-2} c_{m,n-3}. \end{aligned}$$

Using these equalities, prove that $c_{0,0} = -c_{1,-1} = c_{2,-1} = -c_{0,1}$, and that $c_{1,0} = -c_{2,0} = c_{1,1} = -c_{2,1} = c_{2,1} = 0$, and $c_{0,-1} = -c_{0,-1} = 0$. Now choose values of x and y and use the Jacobi triple product identity to prove that

$$c_{0,0} = \prod_{k=1}^{\infty}(1-q^k)^{-2}.$$

2.3 Recursive formulæ

In this section, we shall use Theorem 1.1 to prove the recursive formula for plane partitions that is given in Theorem 1.2. We postpone the proof of Theorem 1.1 until Chapters 3 and 4 where it will provide an introduction to an assortment of techniques for finding plane partition generating functions.

It would be nice to have a result like Euler's pentagonal number theorem for the product

$$\prod_{j=1}^{\infty}(1-q^j)^j.$$

Unfortunately, no such result is known. In the exercises, I invite the reader to use *Mathematica* to search for such a formula. We shall instead use a generic approach that is applicable to many different generating functions. It begins by differentiating each side of equation (1.5).

The product rule for differentiation can be written as

$$\begin{aligned}\frac{d}{dq}\left(\prod_j f_j(q)\right) &= \sum_k \left(\frac{f_k'(q)}{f_k(q)}\prod_j f_j(q)\right) \\ &= \left(\sum_k \frac{f_k'(q)}{f_k(q)}\right)\prod_j f_j(q). \qquad (2.16)\end{aligned}$$

If $f_k(q) = (1-q^k)^{-k}$, then

$$\frac{f_k'(q)}{f_k(q)} = \frac{k^2 q^{k-1}}{1-q^k}.$$

Differentiating each side of equation (1.5) yields

$$\begin{aligned}\sum_{n=1}^{\infty} n\,pp(n)\,q^{n-1} &= \left(\sum_{k=1}^{\infty}\frac{k^2q^{k-1}}{1-q^k}\right)\prod_{j=1}^{\infty}\frac{1}{(1-q^j)^j} \\ &= \left(\sum_{k=1}^{\infty}\frac{k^2q^{k-1}}{1-q^k}\right)\left(\sum_{m=0}^{\infty} pp(m)\,q^m\right), \\ \sum_{n=1}^{\infty} n\,pp(n)\,q^{n} &= \left(\sum_{k=1}^{\infty}\frac{k^2q^{k}}{1-q^k}\right)\left(\sum_{m=0}^{\infty} pp(m)\,q^m\right). \qquad (2.17)\end{aligned}$$

The first sum of the right-hand side can be written as

$$\begin{aligned}\sum_{k=1}^{\infty}\frac{k^2q^k}{1-q^k} &= \sum_{k=1}^{\infty} k^2q^k \sum_{t=0}^{\infty} q^{tk} \\ &= \sum_{k=1}^{\infty}\sum_{t=0}^{\infty} k^2 q^{k(t+1)}. \qquad (2.18)\end{aligned}$$

We now make the substitution $j = k(t+1)$. Once we have chosen the value of j, k can be any positive integer that divides evenly into j (this is written $k|j$, read "k divides j").

$$\begin{aligned}\sum_{k=1}^{\infty}\frac{k^2q^k}{1-q^k} &= \sum_{j=1}^{\infty}\sum_{k|j} k^2\,q^j \\ &= \sum_{j=1}^{\infty}\sigma_2(j)\,q^j, \qquad (2.19)\end{aligned}$$

where $\sigma_2(j)$ is standard shorthand for $\sum_{k|j} k^2$. We substitute this back into equation (2.17). On the right we set $n = j + m$ $(m = n - j)$ so that once n is chosen, j can be any positive integer up to n:

$$\begin{aligned} \sum_{n=1}^{\infty} n\, pp(n)\, q^n &= \left(\sum_{j=1}^{\infty} \sigma_2(j)\, q^j\right)\left(\sum_{m=0}^{\infty} pp(m)\, q^m\right) \\ &= \sum_{n=1}^{\infty}\sum_{j=1}^{n} \sigma_2(j)\, pp(n-j)\, q^n. \end{aligned} \tag{2.20}$$

Theorem 1.2 now follows by comparing the coefficients of q^n on each side.

Counting alternating sign matrices

Mills, Robbins, and Rumsey found the patterns that led them to their formula for the number of alternating sign matrices by actually counting these matrices, but to do their counting efficiently, they used a recursive formula. We begin the process of finding this formula by looking at our alternating sign matrices in a different way.

Given an alternating sign matrix, we first construct a matrix of 0s and 1s in which the entry in the ith row, jth column is the sum of the entries from rows 1 through i of the jth column of the alternating sign matrix. Because of the alternating sign condition, all of these sums must be 0 or 1. We note that we can recover the original alternating sign matrix by taking the difference between each entry and the entry above it. As an example, we have the correspondence

$$\begin{pmatrix} 0 & 1 & 0 & 0 & 0 \\ 0 & 0 & 1 & 0 & 0 \\ 1 & -1 & 0 & 0 & 1 \\ 0 & 1 & -1 & 1 & 0 \\ 0 & 0 & 1 & 0 & 0 \end{pmatrix} \Longleftrightarrow \begin{pmatrix} 0 & 1 & 0 & 0 & 0 \\ 0 & 1 & 1 & 0 & 0 \\ 1 & 0 & 1 & 0 & 1 \\ 1 & 1 & 0 & 1 & 1 \\ 1 & 1 & 1 & 1 & 1 \end{pmatrix}$$

Since each row of an alternating sign matrix has one more 1 than -1, each row of the corresponding matrix has one more 1 than the row above it: row i will contain i 1s and $n - i$ 0s. We now construct a triangular arrangement in which row i records the numbers of the columns that contain a 1:

$$\begin{pmatrix} 0 & 1 & 0 & 0 & 0 \\ 0 & 1 & 1 & 0 & 0 \\ 1 & 0 & 1 & 0 & 1 \\ 1 & 1 & 0 & 1 & 1 \\ 1 & 1 & 1 & 1 & 1 \end{pmatrix} \Longleftrightarrow \begin{array}{ccccccccc} & & & & 2 & & & & \\ & & & 2 & & 3 & & & \\ & & 1 & & 3 & & 5 & & \\ & 1 & & 2 & & 4 & & 5 & \\ 1 & & 2 & & 3 & & 4 & & 5 \end{array}$$

A **monotone triangle** of size n is a triangular arrangement of $n(n+1)/2$ integers taken from the set $\{1, 2, \ldots, n\}$ with the following properties where $a_{ij}, 1 \leq j \leq i$, denotes the jth entry in row i, counted from the top:

1. $a_{ij} < a_{i\,j+1}, \quad 1 \leq j < i,$
2. $a_{ij} \leq a_{i-1\,j} \leq a_{i\,j+1}, \quad 1 \leq j < i.$

An (n, k)-**monotone trapezoid**, $1 \leq k \leq n$, is an arrangement that forms the bottom k rows of a monotone triangle of size n. For example, the following is a $(6, 3)$-monotone trapezoid:

$$\begin{array}{ccccccccccc} & & 1 & & 3 & & 4 & & 6 & & \\ & 1 & & 2 & & 4 & & 5 & & 6 & \\ 1 & & 2 & & 3 & & 4 & & 5 & & 6 \end{array}$$

An (n, n)-monotone trapezoid is just a monotone triangle, and the number of (n, n)-monotone triangles is the number of alternating sign matrices of size n. At the other extreme, there is only one $(n, 1)$-monotone trapezoid:

$$1 \quad 2 \quad 3 \quad \cdots \quad n$$

Our recursive formula will use the number of $(n, k-1)$-monotone trapezoids to find the number of (n, k)-monotone trapezoids.

Given a positive integer n and $S = \{a_1, a_2, \ldots, a_k\}$, a non-empty subset of $\{1, 2, \ldots, n\}$, we define the function $MT_n(S)$ as the number of $(n, n+1-k)$-monotone trapezoids whose top row consists of the integers $a_1, a_2, \ldots, a_k$. The function $MT_n(j) = MT_n(\{j\})$ is the number of monotone triangles with a j at the apex. This is the number of alternating sign matrices in which the first row has a 1 in column j. The total number of alternating sign matrices of size n is the sum $\sum_{j=1}^{n} MT_n(j)$.

We can now set up our recursion; $MT_n(\{a_1, a_2, \ldots, a_k\})$ equals the sum of $MT_n(\{b_1, b_2, \ldots, b_{k+1}\})$ over all possible rows $1 \leq b_1 < b_2 <$

$\cdots < b_{k+1} \le n$ that might sit below $a_1 < a_2 < \cdots < a_k$:

$$MT_n(\{a_1, a_2, \ldots, a_k\}) = \sum_{\substack{b_1 \le a_1 \le b_2 \le a_2 \le b_3 \le \cdots \le a_k \le b_{k+1} \\ 1 \le b_1 < b_2 < b_3 < \cdots < b_{k+1} \le n}} MT_n(\{b_1, b_2, \ldots, b_{k+1}\}). \quad (2.21)$$

For example,

$$\begin{aligned} MT_5(\{1,3,5\}) &= MT_5(\{1,2,3,5\}) + MT_5(\{1,2,4,5\}) \\ &\quad + MT_5(\{1,3,4,5\}) \\ &= MT_5(\{1,2,3,4,5\}) + MT_5(\{1,2,3,4,5\}) \\ &\quad + MT_5(\{1,2,3,4,5\}) \\ &= 3. \end{aligned}$$

In the exercises, there is a *Mathematica* algorithm for implementing this recursion. Each iteration runs quickly, but if n is large there are a lot of intermediate values to be calculated. There are 2^n subsets of $\{1, 2, \ldots, n\}$. The empty set does not occur, and $MT_n(S) = 1$ whenever S contains n or $n-1$ elements, but that still leaves $2^n - n - 2$ values to be calculated and stored. This can be improved, but a recursive approach to finding the number of alternating sign matrices of size n can only be implemented up to about $n = 20$. Fortunately, that is more than far enough to see the patterns that are emerging.

Exercises

2.3.1 If m and n are relatively prime integers, then each positive divisor, k, of mn has a unique representation as $k = k_1 k_2$ where k_1 is a positive divisor of m and k_2 is a positive divisor of n. Use this fact to prove that if m and n are relatively prime then

$$\sigma_2(mn) = \sigma_2(m)\sigma_2(n). \quad (2.22)$$

2.3.2 Is equation (2.22) *ever* true if m and n are *not* relatively prime?

2.3.3 Show that if p is a prime, then

$$\sigma_2(p^a) = 1 + p^2 + p^4 + \cdots + p^{2a} = (p^{2a+2} - 1)/(p^2 - 1),$$

and, therefore,

$$\sigma_2(n) = \prod_{p|n} \frac{p^{2a_p+2} - 1}{p^2 - 1},$$

where the product is over all primes that divide n and a_p is the largest integer for which p^{a_p} divides n.

2.3.4 As in Section 2.1, we let $p_{\mathcal{O}}(n)$ denote the number of partitions of n into odd parts. Using the fact that the generating function for these partitions is given by $\prod(1-q^{2m-1})^{-1}$, prove that

$$p_{\mathcal{O}}(n) = \frac{1}{n}\sum_{j=1}^{n} \sigma_{\text{odd}}(j)\, p_{\mathcal{O}}(n-j),$$

where $\sigma_{\text{odd}}(j)$ is the sum of all odd divisors of j.

2.3.5 As in Section 2.1, we let $p_{\mathcal{D}}(n)$ denote the number of partitions of n into distinct parts. Using the fact that the generating function for these partitions is given by $\prod(1+q^m)$, prove that

$$p_{\mathcal{D}}(n) = \frac{1}{n}\sum_{j=1}^{n} \sigma_{\text{alt}}(j)\, p_{\mathcal{D}}(n-j),$$

where $\sigma_{\text{alt}}(j) = \sum_{k|j} k(-1)^{1+j/k}$.

2.3.6 Give two proofs that for all positive integers j, $\sigma_{\text{odd}}(j) = \sigma_{\text{alt}}(j)$. This can be proved directly or by using the two previous exercises.

2.3.7 Differentiate $(x-x^7)^k/(1-x)^k$, the generating function for $a_k(n)$, the number of ways of rolling k distinct dice to get a total of n (see exercise 2.1.5), to find a recursive formula for $a_k(n)$.

2.3.8 Given the infinite product $\prod_{j=1}^{\infty}(1-q^j)^{-a_j}$, where the a_j are any integers (positive, negative, or zero), use the method for finding the recursive formula for plane partitions to prove that the coefficients in the power series expansion of this product, $1+\sum b_n q^n$, must satisfy

$$nb_n = \sum_{j=1}^{n} D_j b_{n-j}, \quad \text{where } D_j = \sum_{d|j} d a_d. \tag{2.23}$$

The following *Mathematica* code will convert from the sequence $\{a_1,\ldots,a_n\}$ to the sequence $\{1,b_1,\ldots,b_n\}$:

```
A2B[list_]:=Module[{b,d}, d[j_]:=d[j]=
   Apply[Plus, Divisors[j]* list[[Divisors[j]]]];
   b[0]=1;b[n_]:=b[n]=(1/n)Sum[d[j] b[n-j],{j,n}];
   Table[b[n],{n,0,Length[list]}]]
```

2.3.9 Use the recursion in equation (2.23) with $a_d = -d$ to find at least the first ten coefficients in the power series expansion of $\prod_{j=1}^{\infty}(1-q^j)^j$. Is there any apparent pattern?

2.3.10 Show that equation (2.23) implies that

$$na_n = D_n - \sum_{d|n,\ d<n} da_d, \quad \text{where } D_m = mb_m - \sum_{j=1}^{m-1} D_j b_{m-j}. \tag{2.24}$$

The following *Mathematica* code will convert from the sequence $\{b_1, \ldots, b_n\}$ to the sequence $\{a_1, \ldots, a_n\}$:

```
B2A[list_]:= Module[{a,d,alist,pdiv},a[1]=list[[1]];
  d[n_]:=d[n]=n*list[[n]]-Sum[d[j]list[[n-j]],
     {j,n-1}];alist[n_]:=Table[a[j],{j,n}];
  pdiv[n_]:=Delete[Divisors[n],Length[Divisors[n]]];
  a[n_]:=a[n]=(1/n)(d[n] - Apply[Plus,pdiv[n]
     *alist[n-1][[pdiv[n]]]]);
  Table[a[n],{n,Length[list]}]]
```

2.3.11 Prove that every power series with constant term 1 and integer coefficients has a unique representation as an infinite product of the form $\prod(1-q^j)^{-a_j}$ where the a_j are integers. Note that equation (2.24) implies that they must be unique and that ja_j must be an integer.

2.3.12 Use the fact that $j^2 = 1 + 3 + 5 + \cdots + (2j-1)$ to prove that $q^{j^2}/(1-q)(1-q^2)\cdots(1-q^j)$ is the generating function for partitions into exactly j parts that each differ by at least two. It follows that

$$1 + \sum_{j=1}^{\infty} \frac{q^{j^2}}{(1-q)(1-q^2)\cdots(1-q^j)}$$

is the generating function for partitions into parts that differ by at least two. For example, the partitions of 10 into parts that differ by at least two are 10, 9+1, 8+2, 7+3, 6+4, 6+3+1. How many such partitions are there of 12? The following *Mathematica* command will generate a list of the coefficients of q through q^n in the power series expansion of this generating function,

```
RR1[n_]:=Drop[CoefficientList[Normal[Series[
   1+Sum[q^(j^2)/Product[1-q^i,{i,j}],
   {j,Floor[Sqrt[n]]}],{q,0,n}]],q],1]
```

Use `B2A[RR1[100]]` to find the first hundred values of a_j. There is a pattern here that has been independently observed and then proven by many of the important players in our story including L. J. Rogers (1894), I. J. Schur (1917), S. Ramanujan (1919), and R. J. Baxter (1981). What is the identity that they found? It is known as the first **Rogers–Ramanujan identity**. We shall see a proof of it in the exercises to Section 5.2.

2.3.13 Explain why

$$1+\sum_{j=1}^{\infty} \frac{q^{j^2+j}}{(1-q)(1-q^2)\cdots(1-q^j)}$$

is the generating function for partitions into parts that differ by at least two and are greater than or equal to 2. Find the infinite product that should be equal to this sum, what is known as the second Rogers–Ramanujan identity.

2.3.14 Find the alternating sign matrices that correspond to the following monotone triangles:

a

$$\begin{array}{ccccccc} & & & 3 & & & \\ & & 1 & & 4 & & \\ & 1 & & 2 & & 4 & \\ 1 & & 2 & & 3 & & 4 \end{array}$$

b

$$\begin{array}{ccccccccccc} & & & & & 3 & & & & & \\ & & & & 2 & & 4 & & & & \\ & & & 1 & & 3 & & 5 & & & \\ & & 1 & & 2 & & 4 & & 6 & & \\ & 1 & & 2 & & 3 & & 4 & & 6 & \\ 1 & & 2 & & 3 & & 4 & & 5 & & 6 \end{array}$$

2.3.15 Given a monotone triangle, how do you recognize how many -1s are in the corresponding alternating sign matrix?

2.3.16 Define the set of monotone triangles that correspond to permutation matrices (alternating sign matrices without any -1s).

2.3.17 Explain why $MT_n(S) = 1$ whenever S contains n or $n-1$ elements.

2.3.18 The following *Mathematica* commands allow you to calculate `MT[n][S]` $= MT_n(S)$ and `ASM[n]`, the total number of $n \times n$ alternating sign matrices. Explore the values of $MT_n(\{1,i\})$, $1 < i \leq n$, and conjecture a formula for these values.

```
nextlist[n_,alist_] := Select[
  Distribute[Apply[Range,
  Partition[Prepend[Append[alist,n],1], 2, 1], 1],
    List], # == Union[#] &];
MT[n_][blist_]:=MT[n][blist]=If[Length[blist]>=n-1,
    1,Plus@@Map[MT[n],nextlist[n,blist]],{1}];
ASM[n_]:=Sum[MT[n][{i}],{i,n}]
```

2.3.19 Prove the conjecture you made in exercise 2.3.18.

2.4 Determinants

It should come as no surprise that determinants are caught up in the web spun by the theory of partitions. They are, after all, summations over the symmetric group. Their evaluation can often be accomplished by exploiting this link to groups of symmetries. A fundamental determinant that will reappear throughout our story is the Vandermonde determinant, named for Alexandre Théophile Vandermonde (1735–1796):

$$\prod_{1\le i<j\le n} (x_i - x_j) = \begin{vmatrix} x_1^{n-1} & x_2^{n-1} & \dots & x_n^{n-1} \\ x_1^{n-2} & x_2^{n-2} & \dots & x_n^{n-2} \\ \vdots & \vdots & & \vdots \\ x_1 & x_2 & \dots & x_n \\ 1 & 1 & \dots & 1 \end{vmatrix}, \tag{2.25}$$

$$= \sum_{\sigma\in\mathcal{S}_n} (-1)^{\mathcal{I}(\sigma)} \prod_{i=1}^{n} x_{\sigma(i)}^{n-i}, \tag{2.26}$$

where $\mathcal{I}(\sigma)$ denotes the **inversion number** of the permutation σ, the minimal number of transpositions of adjacent elements needed to get from the identity permutation to σ. On occasion, it will be useful to reverse the order of the rows and to write this identity as

$$\prod_{1\le i<j\le n} (x_j - x_i) = \left|x_j^{i-1}\right| = \sum_{\sigma\in\mathcal{S}_n} (-1)^{I(\sigma)} \prod_{i=1}^{n} x_{\sigma(i)}^{i-1}.$$

In 1772, Vandermonde published "Mémoire sur l'élimination," outlining a general method for solving systems of linear equations. While earlier mathematicians including Leibniz, Maclaurin, Cramer, and Euler were aware of determinants as they are used in finding the solution of a system of linear equations, Vandermonde's treatise was a benchmark in

Figure 2.4. Augustin-Louis Cauchy (From the American Institute of Physics Emilio Segrè Visual Archives).

the clarification of rules for the evaluation of a determinant of arbitrary size. Vandermonde, however, did not discover the general product identity that carries his name. This was found by Augustin-Louis Cauchy in 1812 (Cauchy 1815).

Cauchy was then a young lieutenant in Cherbourg working on preparations for the invasion of England and writing mathematics in his spare time. He was actually studying **alternating functions**. These are functions that change sign when we transpose any two of the variables. The title of Cauchy's paper defines alternating functions: "Memoir on functions that can only take on two equal values of opposite sign when they are acted on by transpositions of two of their variables" (1815).

In this important paper, Cauchy gives general rules for evaluating determinants, becomes the first person to use the term "determinant" in our modern sense, and gives the first proof† that the determinant of a product of two matrices is the product of the individual determinants. Cauchy's first definition of the determinant is expressed in terms of the

† Binet presented a different proof of this same fact on the same day, 30 November, 1812. Both men agreed that their proofs had been discovered independently.

Vandermonde product. To find the determinant of

$$\begin{pmatrix} x_{0,0} & x_{0,1} & \cdots & x_{0,n-1} \\ x_{1,0} & x_{1,1} & \cdots & x_{1,n-1} \\ \vdots & \vdots & & \vdots \\ x_{n-1,0} & x_{n-1,1} & \cdots & x_{n-1,n-1} \end{pmatrix},$$

you expand the product

$$\prod_{0 \le i < j \le n-1} (x_j - x_i),$$

and then replace each x_i^k by $x_{i,k}$. For example,

$$\begin{aligned} (x_1 - x_0)(x_2 - x_0)(x_2 - x_1) &= x_0^0 x_1^1 x_2^2 + x_0^1 x_1^2 x_2^0 + x_0^2 x_1^0 x_2^1 \\ &\quad - x_0^1 x_1^0 x_2^2 - x_0^2 x_1^1 x_2^0 - x_0^0 x_1^2 x_2^1, \end{aligned}$$

and therefore,

$$\begin{vmatrix} x_{0,0} & x_{0,1} & x_{0,2} \\ x_{1,0} & x_{1,1} & x_{1,2} \\ x_{2,0} & x_{2,1} & x_{2,2} \end{vmatrix} = \begin{aligned} & x_{0,0}x_{1,1}x_{2,2} + x_{0,1}x_{1,2}x_{2,0} + x_{0,2}x_{1,0}x_{2,1} \\ & - x_{0,1}x_{1,0}x_{2,2} - x_{0,2}x_{1,1}x_{2,0} - x_{0,0}x_{1,2}x_{2,1}. \end{aligned}$$

Cauchy later shows that this definition is equivalent to the one with which we are familiar, expressing the determinant as a summation over permutations.

Proof by exploitation of symmetry

Our proof of the Vandermonde determinant formula, as well as the proof of Proposition 2.8, comes directly from Cauchy's 1812 paper. It shows that, in some sense, this is not a very deep or difficult result. If we transpose any two columns of (x_j^{n-i}), we change the sign of the determinant. It follows that the determinant of this matrix is an alternating polynomial of degree $n(n-1)/2$ in the variables $x_1, x_2, \ldots, x_n$. The Vandermonde identity is now a simple consequence of the following proposition which will be an important tool for investigation in the remaining chapters.

Proposition 2.8 *If $f(x_1, x_2, \ldots, x_n)$ is an alternating polynomial of degree d, then*

$$\frac{f(x_1, x_2, \ldots, x_n)}{\prod_{1 \le i < j \le n} (x_i - x_j)}$$

is a symmetric polynomial of degree $d - n(n-1)/2$.

This proposition implies that $\det(x_j^{n-i})/\prod(x_i - x_j)$ is a constant:

$$\det(x_j^{n-i}) = C \prod_{1\le i<j\le n} (x_i - x_j).$$

To see that $C = 1$, we compare the coefficients of $x_1^{n-1}x_2^{n-2}\cdots x_{n-1}$ on each side.

Proof: This proposition says three things about $f/\prod(x_i - x_j)$: that it is symmetric, that it is a polynomial, and that it has degree $d - n(n-1)/2$. Since $\prod_{1\le i<j\le n}(x_i - x_j)$ is itself an alternating polynomial of degree $n(n-1)/2$, the only piece that needs to be proven is that our ratio is still a polynomial.

We view f as a polynomial in x_1 with coefficients expressed in terms of the other variables. Since interchanging the first two variables changes the sign, we have that

$$f(x_2, x_2, \ldots, x_n) = -f(x_2, x_2, \ldots, x_n),$$

which can only happen if $f(x_2, x_2, \ldots, x_n) = 0$. Similarly, our polynomial is 0 when $x_1 = x_3, x_4, \ldots, x_n$, and so we have identified $n-1$ distinct roots:

$$f(x_1, x_2, \ldots, x_n) = g(x_1, x_2, \ldots, x_n) \prod_{j=2}^{n} (x_1 - x_j)$$

where g is a polynomial in $x_1, x_2, \ldots, x_n$. If $n = 2$, then

$$g(x_1, x_2) = \frac{f(x_1, x_2)}{x_1 - x_2}$$

is a symmetric polynomial in x_1 and x_2.

If n is larger than 2, then we can proceed by induction on the number of variables. We consider $g(x_1, x_2, \ldots, x_n)$ as a polynomial in $x_2, \ldots, x_n$ with coefficients expressed in terms of x_1. This is an alternating polynomial in the $n-1$ variables $x_2, \ldots, x_n$. By the induction hypothesis, $g(x_1, x_2, \ldots, x_n)/\prod_{2\le i<j\le n}(x_i - x_j)$ is a symmetric polynomial in $x_2, \ldots, x_n$. Since the denominator does not involve x_1, it is still a polynomial in x_1. Therefore, we have that

$$\frac{f(x_1, x_2, \ldots, x_n)}{\prod_{1\le i<j\le n}(x_i - x_j)} = \frac{g(x_1, x_2, \ldots, x_n)}{\prod_{2\le i<j\le n}(x_i - x_j)}$$

is a polynomial in $x_1, x_2, \ldots, x_n$.

Q.E.D.

Krattenthaler's formula

In 1990, Christian Krattenthaler published the following extension of the Vandermonde formula. We shall use it in our first proof of the generating function for plane partitions.

Theorem 2.9 (Krattenthaler's formula) *Given arbitrary values for* $x_1, \ldots, x_n$, $a_2, \ldots, a_n$, *and* $b_2, \ldots, b_n$, *we have that*

$$\det\left((x_i + a_n)\cdots(x_i + a_{j+1})(x_i + b_j)\cdots(x_i + b_2)\right)_{i,j=1}^{n} = \prod_{1 \le i < j \le n} (x_i - x_j) \prod_{2 \le i \le j \le n} (b_i - a_j). \tag{2.27}$$

Proof: The determinant on the left side of equation (2.27) is an alternating function of $x_1, \ldots, x_n$ and so is equal to the Vandermonde product times a symmetric polynomial in $x_1, \ldots, x_n$. As a function of x_1, the determinant is a polynomial of degree $n-1$. The Vandermonde product is also a polynomial of degree $n-1$ in x_1, and therefore this unknown symmetric polynomial must be constant with respect to x_1. Since it is symmetric in the xs, it must be constant with respect to every other x_i. We have established that

$$\det\left((x_i + a_n)\cdots(x_i + a_{j+1})(x_i + b_j)\cdots(x_i + b_2)\right)_{i,j=1}^{n} = C \prod_{1 \le i < j \le n} (x_i - x_j),$$

where C may depend on the a_i and b_i but is independent of the x_i.

To find C, we set $x_i = -a_i$ for $2 \le i \le n$. All of the entries in our matrix below the main diagonal become 0, and the determinant is just the product of entries on the main diagonal:

$$\begin{aligned}(x_1 + a_n)&\cdots(x_1 + a_2)\\ &\times \prod_{i=2}^{n}(-a_i + a_n)\cdots(-a_i + a_{i+1})(-a_i + b_i)\cdots(-a_i + b_2)\\ &= C(x_1 + a_2)\cdots(x_1 + a_n) \prod_{2 \le i < j \le n} (-a_i + a_j).\end{aligned}$$

Comparing both sides of this equality, we see that C is the desired product,

$$C = \prod_{2 \le i \le j \le n} (b_i - a_j).$$

Q.E.D.

The Weyl denominator formula

The Vandermonde determinant identity is a special case of a general determinant evaluation formula named for the great geometer and algebraist Hermann Weyl (1885–1955). Weyl used it to simplify the denominator of a particular formula, which is how it got its name. We shall need another special case of this formula in Chapter 4 when we prove MacMahon's conjectured generating function for symmetric plane partitions.

The Weyl formula is expressed in terms of **root systems**, sets of vectors in $\mathbf{R}^n$ that satisfy certain symmetry conditions. The Vandermonde formula corresponds to the set $\mathbf{A_n} = \{\pm(\vec{e_i} - \vec{e_j}) \mid 1 \le i < j \le n+1\}$, where $\vec{e_i}$ is the unit vector in the ith direction. (Note that despite appearances, this set lies in an n-dimensional space since the sum of the coordinates of any point in this space is 0.) The other infinite families of root systems are

$$\begin{aligned}
\mathbf{D_n} &= \mathbf{A_{n-1}} \cup \{\pm(\vec{e_i} + \vec{e_j}) \mid 1 \le i < j \le n\}, \\
\mathbf{B_n} &= \mathbf{D_n} \cup \{\pm\vec{e_i} \mid 1 \le i \le n\}, \\
\mathbf{C_n} &= \mathbf{D_n} \cup \{\pm 2\vec{e_i} \mid 1 \le i \le n\}.
\end{aligned}$$

We shall state the specific form of the Weyl denominator formula for each of these three families of root systems. Those interested in the general statement of the formula can find it in Carter's *Simple Groups of Lie Type* (1989, chap. 10).

Theorem 2.10 (The $\mathbf{B_n}$, $\mathbf{C_n}$, and $\mathbf{D_n}$ Weyl denominator formulas) *The first equality restates the determinant as a sum over permutations and subsets $S \subseteq \{1, 2, \ldots, n\}$ for which i is in S if and only if we choose the second term of the binomial in column i, row $\sigma(i)$. The Weyl denominator formula is given in the second equality which expresses the determinant as a product.*

$\mathbf{B_n}$:

$$\det\begin{pmatrix}
1 - x_1^{2n-1} & 1 - x_2^{2n-1} & \cdots & 1 - x_n^{2n-1} \\
x_1 - x_1^{2n-2} & x_2 - x_2^{2n-2} & \cdots & x_n - x_n^{2n-2} \\
\vdots & \vdots & & \vdots \\
x_1^{n-1} - x_1^n & x_2^{n-1} - x_2^n & \cdots & x_n^{n-1} - x_n^n
\end{pmatrix}$$

$$= \sum_{\substack{\sigma \in S_n \\ S \subseteq \{1,\ldots,n\}}} (-1)^{\mathcal{I}(\sigma)+|S|} \prod_{i \in S} x_i^{2n-\sigma(i)} \prod_{i \notin S} x_i^{\sigma(i)-1}$$

$$= \left(\prod_{i=1}^{n}(1-x_i)\right)\left(\prod_{1\le i<j\le n}(x_i-x_j)(x_ix_j-1)\right), \qquad (2.28)$$

$\mathbf{C_n}$:

$$\det\begin{pmatrix} 1-x_1^{2n} & 1-x_2^{2n} & \cdots & 1-x_n^{2n} \\ x_1-x_1^{2n-1} & x_2-x_2^{2n-1} & \cdots & x_n-x_n^{2n-1} \\ \vdots & \vdots & & \vdots \\ x_1^{n-1}-x_1^{n+1} & x_2^{n-1}-x_2^{n+1} & \cdots & x_n^{n-1}-x_n^{n+1} \end{pmatrix}$$

$$= \sum_{\substack{\sigma\in S_n\\ S\subseteq\{1,\ldots,n\}}} (-1)^{\mathcal{I}(\sigma)+|S|}\prod_{i\in S}x_i^{2n+1-\sigma(i)}\prod_{i\notin S}x_i^{\sigma(i)-1}$$

$$= \left(\prod_{i=1}^{n}(1-x_i^2)\right)\left(\prod_{1\le i<j\le n}(x_i-x_j)(x_ix_j-1)\right), \qquad (2.29)$$

$\mathbf{D_n}$:

$$\det\begin{pmatrix} 1+x_1^{2n-2} & 1+x_2^{2n-2} & \cdots & 1+x_n^{2n-2} \\ x_1+x_1^{2n-3} & x_2+x_2^{2n-3} & \cdots & x_n+x_n^{2n-3} \\ \vdots & \vdots & & \vdots \\ x_1^{n-1}+x_1^{n-1} & x_2^{n-1}+x_2^{n-1} & \cdots & x_n^{n-1}+x_n^{n-1} \end{pmatrix}$$

$$= \sum_{\substack{\sigma\in S_n\\ S\subseteq\{1,\ldots,n\}}} (-1)^{\mathcal{I}(\sigma)}\prod_{i\in S}x_i^{2n-1-\sigma(i)}\prod_{i\notin S}x_i^{\sigma(i)-1}$$

$$= 2\left(\prod_{1\le i<j\le n}(x_i-x_j)(x_ix_j-1)\right). \qquad (2.30)$$

Proof: We shall prove only the formula for $\mathbf{B_n}$. The others are proven similarly. The first equality follows from first writing the determinant as a sum over permutations and then expanding each of the binomials (i is an element of S if and only if we choose the second term from the binomial). The substance of this theorem is in the second equality.

Let $G_n(x_1, ...x_n)$ denote the determinant in equation (2.28). We shall prove the second equality by induction on n. It is easy to verify that this equality is correct for $n = 1$ or 2. We assume that

$$G_{n-1}(x_2,\ldots,x_n) = \left(\prod_{i=2}^{n}(1-x_i)\right)\left(\prod_{2\le i<j\le n}(x_i-x_j)(x_ix_j-1)\right).$$

When $x_1 = 1$, each entry in the first column is 0. When $x_1 = x_j$ for any j larger than 1, the first and jth columns are identical. When $x_1 = x_j^{-1}$ for any j larger than 1, then the first column is x_j^{-2n+1} times the jth column. In each of these cases, the determinant is zero. If we view our determinant, $G_n(x_1, \ldots, x_n)$, as a polynomial of degree $2n-1$ in x_1, we see that it has roots at $1, x_2, .., x_n, x_2^{-1}, ..., x_n^{-1}$. This tells us that

$$\begin{aligned} G_n(x_1, \ldots, x_n) &= c(x_1 - 1) \prod_{j=2}^{n} (x_1 - x_j)(x_1 - x_j^{-1}) \\ &= -c(x_2 \cdots x_n)^{-1}(1 - x_1) \prod_{j=2}^{n} (x_1 - x_j)(x_1 x_j - 1), \end{aligned}$$

where c does not depend on x_1, though it may depend on the other variables. The coefficient of x_1^{2n-1} on the right side of this equation is c. In the determinant, the coefficient of x_1^{2n-1} is

$$-\det \begin{pmatrix} x_2 - x_2^{2n-2} & \cdots & x_n - x_n^{2n-2} \\ \vdots & & \vdots \\ x_2^{n-1} - x_2^{n} & \cdots & x_n^{n-1} - x_n^{n} \end{pmatrix} = -x_2 \cdots x_n G_{n-1}(x_2, ..., x_n).$$

Therefore, c equals $-x_2 \cdots x_n G_{n-1}(x_2, ..., x_n)$.

Q.E.D.

Exercises

2.4.1 Exercises 2.4.1 to 2.4.6 lead through a combinatorial proof of the Vandermonde determinant formula that is based on one published by Gessel (1979). A **tournament** on n vertices is a complete, directed graph with n labeled vertices. Let $\mathcal{T}_n$ be the set of all tournaments on n vertices. Given a tournament $T \in \mathcal{T}_n$, let $U(T)$ be the number of **upsets**, defined to be the number of edges directed from the vertex with the larger label to the vertex with the smaller label. Let $\omega(i)$ be the **out-degree** of vertex i, the number of directed edges going out from i. Show that the Vandermonde product can be expressed as the following sum over tournaments:

$$\prod_{1 \le i < j \le n} (x_i - x_j) = \sum_{T \in \mathcal{T}_n} (-1)^{U(T)} \prod_{i=1}^{n} x_i^{\omega(i)}. \qquad (2.31)$$

2.4.2 A tournament is **transitive** if it does not contain any cycles. In other words, in a transitive tournament, if i beats j and j beats k, then i must beat k. Find a natural one-to-one correspondence between transitive tournaments on n vertices and permutations of n letters.

2.4.3 Prove that the number of upsets in a transitive tournament is equal to the inversion number of the corresponding permutation. Show that this implies that the determinant side of the Vandermonde identity can be written as a sum over all transitive tournaments, denoted by $\mathcal{T}_n^T$:

$$\sum_{T \in \mathcal{T}_n^T} (-1)^{U(T)} \prod_{i=1}^{n} x_i^{\omega(i)}.$$

The identity that we want to prove has been shown to be equivalent to

$$\sum_{T \in \mathcal{T}_n} (-1)^{U(T)} \prod_{i=1}^{n} x_i^{\omega(i)} = \sum_{T \in \mathcal{T}_n^T} (-1)^{U(T)} \prod_{i=1}^{n} x_i^{\omega(i)}.$$

2.4.4 Let $\mathcal{T}_n^{NT}$ denote the set of tournaments on n vertices that are not transitive. As we saw in the last exercise, the Vandermonde determinant formula is equivalent to

$$\sum_{T \in \mathcal{T}_n^{NT}} (-1)^{U(T)} \prod_{i=1}^{n} x_i^{\omega(i)} = 0.$$

Explain why this will be proven if we can find a bijection f from $\mathcal{T}_n^{NT}$ to itself such that $f(f(T)) = T$, each vertex has the same out-degree in T as in $f(T)$, and $U(T)$ and $U(f(T))$ have opposite parity.

2.4.5 Prove that in a non-transitive tournament, there must be at least two players with the same number of wins. Hint: The number of wins for each player must be at least 0 and at most $n-1$. Show that if each player has a different number of wins, then the tournament must be transitive.

2.4.6 Prove that the following function f satisfies the criteria of exercise 2.4.4 and thus concludes the proof of the Vandermonde identity. Given a non-transitive tournament, T, define $f(T)$ as the tournament that you obtain when you switch the labels on vertices i and j where (i, j) is the smallest pair of integers for which $\omega(i) = \omega(j)$.

2.4.7 Evaluate $\det\left(\binom{x_i+j}{j}\right)_{i,j=0}^{n}$.

2.4.8 Prove the Weyl denominator formula for the roots systems $\mathbf{C_n}$ and $\mathbf{D_n}$.

2.4.9 Prove that

$$\det\left(\frac{1}{1-x_iy_j}\right)_{i,j=1}^{n} = \prod_{1<i\leq j<n}[(x_i-x_j)(y_i-y_j)]\prod_{i,j=1}^{n}(1-x_iy_j)^{-1}. \quad (2.32)$$

Hint: Use the fact that $\prod(1-x_iy_j)\det(1/(1-x_iy_j))$ is an alternating polynomial in the x_i and also in the y_j, and it is of degree $n-1$ in each of the variables.

2.4.10 Prove Kuperberg's determinant evaluation (1996b):

$$\det\left(\frac{1-s^{i+j-1}}{1-t^{i+j-1}}\right)_{i,j=1}^{n} = t^{n^3/3-n^2/2+n/6}\prod_{1\leq i<j\leq n}(1-t^{j-i})^2\prod_{i,j=1}^{n}\frac{1-s\,t^{j-i}}{1-t^{i+j-1}}. \quad (2.33)$$

Hint: First show that the determinant is a polynomial of degree at most n^2 in s. Next identify the roots. This polynomial has a root of order at least n at $s=1$. For $1\leq k<n$, show that each row vector of the matrix $\left((1-t^{k(i+j-1)})/(1-t^{i+j-1})\right)$ is a linear combination of the vectors

$$\begin{array}{l}(1,1,1,\ldots,1)\\ (1,t,t^2,\ldots,t^{n-1})\\ (1,t^2,t^4,\ldots,t^{2(n-1)})\\ \vdots\\ (1,t^{k-1},t^{2k-2},\ldots,t^{(k-1)(n-1)}),\end{array}$$

and therefore this matrix has rank at most k. It follows that $s=t^k$ is a root of order at least $n-k$. Similarly, there is a root of order at least $n-k$ at $s=t^{-k}$. Finally, use equation (2.32) to prove that the constant term in this polynomial is

$$t^{n^3/3-n^2/2+n/6}\prod_{1\leq i<j\leq n}(1-t^{j-i})^2\prod_{i,j=1}^{n}1/(1-t^{i+j-1}).$$

3

Lattice Paths and Plane Partitions

Gessel and Viennot have developed a powerful technique for enumerating various classes of plane partitions [1985]. There are two fundamental ideas behind this technique. The first is the observation that most classes of plane partitions that are of interest – either by association with the representation theory of the classical groups, or for purely combinatorial reasons – can be interpreted as configurations of nonintersecting paths in a digraph (usually the lattice $\mathbf{Z}^2$). The second is the observation that the number of r-tuples of nonintersecting paths between two sets of r vertices can (often) be expressed as a determinant.

– John Stembridge (1990)

Partitions of integers have been studied extensively because they lie at the crossroads of several mathematical disciplines including symmetric functions and modern algebra as well as elliptic integrals and the theory of modular forms that grew from them. They also appear as soon as one begins to probe the structure of the binomial coefficient,

$$\binom{m+n}{m} = \frac{(m+n)!}{m!\,n!},$$

one of the building blocks of combinatorics. This quantity expresses the number of ways of choosing m objects from a set of $m+n$ objects. Following Richard Guy, it is read as "$m+n$ choose m." There are many equivalent ways of interpreting what the binomial coefficient counts. We shall focus on lattice paths.

How many ways can one travel from the origin, $(0,0)$, to the point (n,m) if m and n are non-negative integers and if each step is one unit either up (north) or to the right (east)? We take a total of $m+n$ steps. Exactly m of these steps must go north. And we can take a step north at any time. From the set of $m+n$ steps, we can choose any m of them to be the steps north, so the answer is

$$\binom{m+n}{m}.$$

The lattice path from $(0,0)$ to $(6,5)$ shown below corresponds to choosing the first, fourth, sixth, ninth, and tenth steps as the steps to the north.

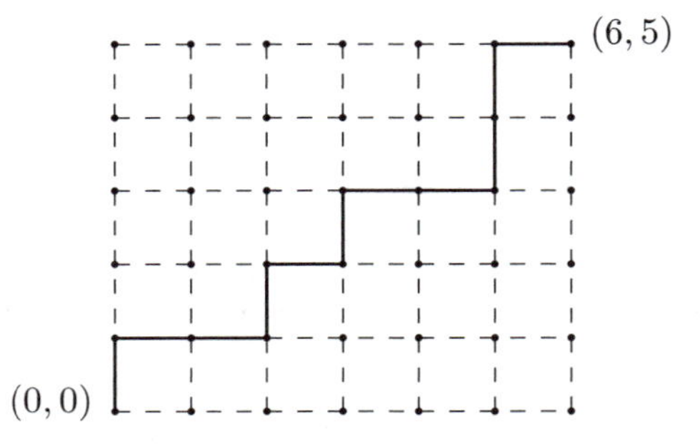

We can record our steps as a sequence of 1s and 0s, 0 for a step to the north and 1 for a step to the east. Our sample lattice path corresponds to the sequence

$$01101011001.$$

3.1 Lattice paths

There are thirty squares in the rectangle that delimits possible lattice paths from $(0,0)$ to $(6,5)$. Each lattice path splits these into two subsets: those to the northwest and those to the southeast. If we put a dot inside each square to the northwest of the lattice path, we get the Ferrers graph of a partition. Our sample path corresponds to the partition $5+5+3+2$:

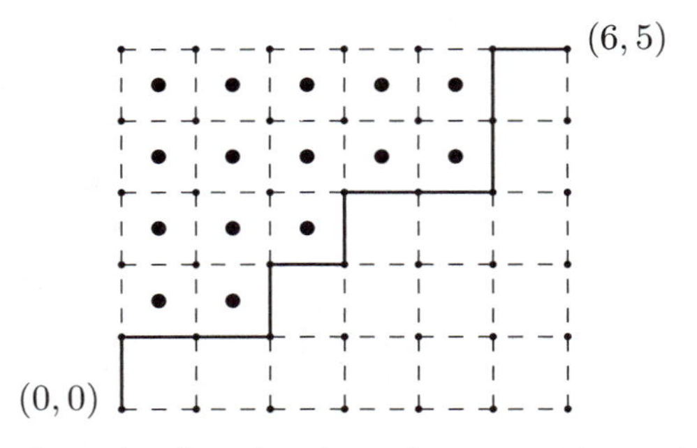

Lattice paths from $(0,0)$ to $(6,5)$ produce partitions with at most five parts, and each part will be less than or equal to six. The correspondence is one-to-one; each partition into at most five parts with each part less than or equal to six defines a unique lattice path from $(0,0)$ to $(6,5)$. We have established the following result.

Proposition 3.1 *The total number of partitions into at most m parts with each part less than or equal to n is equal to* $\binom{m+n}{m}$.

This proposition suggests a generalization of the binomial coefficient. Instead of counting the total number of partitions into at most m parts with each part less than or equal to n, we shall look at the generating function for such partitions. Since we cannot have more than mn dots in the Ferrers graph, the coefficient will be 0 when the power of q is greater than mn. The coefficient of q^{mn} is 1: There is exactly one partition of mn that will fit inside this rectangle. This means that the generating function is a monic polynomial of degree mn.

For $m = 5$, $n = 6$, the polynomial is

$$\begin{aligned} f_{6,5}(q) &= 1 + q + 2q^2 + 3q^3 + 5q^4 + 7q^5 + 10q^6 + 12q^7 + 16q^8 \\ &\quad + 19q^9 + 23q^{10} + 25q^{11} + 29q^{12} + 30q^{13} + 32q^{14} + 32q^{15} \\ &\quad + 32q^{16} + 30q^{17} + 29q^{18} + 25q^{19} + 23q^{20} + 19q^{21} + 16q^{22} \\ &\quad + 12q^{23} + 10q^{24} + 7q^{25} + 5q^{26} + 3q^{27} + 2q^{28} + q^{29} + q^{30}. \end{aligned}$$

There are two easy observations to be made about this particular polynomial which turn out to be true for all such generalizations of binomial coefficients. First, they are **palindromic**: The coefficients read the same forward or backward. This is not hard to prove, and exercises 3.1.11 and 3.1.12 lead to two different proofs. The second observation is that they are **unimodal**, that the coefficients are weakly increasing when less than $mn/2$, and weakly decreasing thereafter. This was first proven by Sylvester (1878). It is surprisingly difficult to prove.

If we set $q = 1$ in the generating function, $f_{m,n}(q)$, then we are adding up all of the coefficients, which means finding the total number of partitions that fit inside our $m \times n$ rectangle. We know that this number is $\binom{m+n}{m}$:

$$f_{m,n}(1) = \binom{m+n}{m}.$$

This means that our generating function is a generalization of the binomial coefficient, and for this reason we use a notation that suggests this similarity,

$$\begin{bmatrix} m+n \\ m \end{bmatrix}_q \quad \text{instead of} \quad f_{m,n}(q).$$

It is common to drop the subscript q and just write $\left[{m+n \atop m}\right]$ unless the variable is something other than q.

The Gaussian polynomial

Our polynomial generating function is called the **Gaussian polynomial** because it was first described and used by C. F. Gauss. He needed it for his evaluation of the **Gaussian sum**:

$$\sum_{j=0}^{k-1} \alpha^{j^2}, \quad \alpha = e^{2\pi ih/k},$$

where k is a positive odd integer and h is relatively prime to k (Rademacher 1964, chap. 11). Gauss discovered that these polynomials could be written as a ratio of products in a manner that parallels the formula for the binomial coefficient as a ratio of factorials.

Proposition 3.2 *If m and n are positive integers, then*

$$\begin{aligned}\begin{bmatrix} m+n \\ m \end{bmatrix}_q &= \frac{(1-q)(1-q^2)\cdots(1-q^{m+n})}{(1-q)(1-q^2)\cdots(1-q^m)(1-q)(1-q^2)\cdots(1-q^n)} \\ &= \prod_{i=1}^{m} \frac{1-q^{n+i}}{1-q^i}. \end{aligned} \tag{3.1}$$

It is often convenient to employ the notation

$$(q;q)_m = (1-q)(1-q^2)\cdots(1-q^m).$$

In this notation, equation (3.1) is written as

$$\begin{bmatrix} m+n \\ m \end{bmatrix}_q = \frac{(q;q)_{m+n}}{(q;q)_m\,(q;q)_n}.$$

Note that this is analogous to the representation of the binomial coefficient as $\binom{m+n}{m} = (m+n)!/m!\,n!$, and it immediately implies that

$$\begin{bmatrix} m+n \\ m \end{bmatrix}_q = \begin{bmatrix} m+n \\ n \end{bmatrix}_q.$$

Proof: This proof will illustrate a general technique of combinatorial arguments, an approach that generalizes proof by induction. We verify that each side of the equation to be proven, equation (3.1), satisfies the same recursive formula and the same boundary conditions. Since each side is uniquely determined by the boundary values and the recursive formula, the two sides must be equal.

For our problem, the boundary values are at $m = 0$ and at $n = 0$. In either case, there is exactly one lattice path from $(0, 0)$ to (n, m); all steps are to the north if $n = 0$; all steps are to the east if $m = 0$. The only possible partition is the empty partition:

$$\begin{bmatrix} m \\ 0 \end{bmatrix} = \begin{bmatrix} m \\ m \end{bmatrix} = 1. \tag{3.2}$$

The binomial coefficient is completely determined by these boundary conditions, $\binom{m}{0} = \binom{m}{m} = 1$, together with a recursive formula,

$$\binom{m+n}{m} = \binom{m+n-1}{m} + \binom{m+n-1}{m-1}, \qquad m, n > 0.$$

There is a similar recursive formula for Gaussian polynomials. Given a partition into at most m parts, each less than or equal to n, there are two mutually exclusive possibilities:

1. All of the parts are strictly less than n; the generating function for these partitions is

$$\begin{bmatrix} m+n-1 \\ m \end{bmatrix}.$$

2. The largest part is exactly n; we record this part with q^n and then observe that what remains is a partition into at most $m - 1$ parts, each less than or equal to n; the generating function for these partitions is

$$q^n \begin{bmatrix} m+n-1 \\ m-1 \end{bmatrix}.$$

This establishes the recursive formula

$$\begin{bmatrix} m+n \\ m \end{bmatrix} = \begin{bmatrix} m+n-1 \\ m \end{bmatrix} + q^n \begin{bmatrix} m+n-1 \\ m-1 \end{bmatrix}. \tag{3.3}$$

The Gaussian polynomial is uniquely defined by the boundary conditions of equation (3.2) and the recursive formula in equation (3.3). We define the product in equation (3.1) to be 1 when $m = 0$. This product is also equal to 1 when $n = 0$. If it satisfies the same recursive formula, then it must equal the Gaussian polynomial. It is easy to check that it does:

$$\prod_{i=1}^{m} \frac{1-q^{n+i}}{1-q^i} = \prod_{i=1}^{m} \frac{1-q^{n-1+i}}{1-q^i} + q^n \prod_{i=1}^{m-1} \frac{1-q^{n+i}}{1-q^i}. \tag{3.4}$$

Q.E.D.

Proposition 3.2 may not look impressive. In fact, it is powerful. It provides a combinatorial interpretation for the coefficients of the polynomial $(q;q)_{m+n}/(q;q)_m(q;q)_n$. Whenever we have such a concrete grasp of any polynomial or power series, we have a depth of understanding of the function that can lead us to new insights. In this particular case, Proposition 3.2 implies a polynomial identity for which the Jacobi triple product identity is a limiting case.

Another proof of Jacobi's triple product identity

The Gaussian polynomial is also known as the q-**binomial coefficient** because of its role in a generalization of the binomial theorem,

$$(1+x)^n = \sum_{i=0}^{n} \binom{n}{i} x^i.$$

Theorem 3.3 The q-binomial theorem *For any positive integer n,*

$$(1+xq)(1+xq^2)\cdots(1+xq^n) = \sum_{i=0}^{n} \begin{bmatrix} n \\ i \end{bmatrix} q^{i(i+1)/2} x^i. \tag{3.5}$$

Proof: If we expand the product $(1+xq)(1+xq^2)\cdots(1+xq^n)$, we get a polynomial in x in which the coefficient of x^i is a polynomial in q:

$$(1+xq)(1+xq^2)\cdots(1+xq^n) = \sum_{i=0}^{n} a_i(q)\, x^i.$$

For example,

$$\begin{aligned} &(1+xq)(1+xq^2)(1+xq^3)(1+xq^4) \\ &\quad = 1 + (q+q^2+q^3+q^4)x + (q^3+q^4+2q^5+q^6+q^7)x^2 \\ &\qquad + (q^6+q^7+q^8+q^9)x^3 + q^{10}x^4 \\ &\quad = 1 + \begin{bmatrix} 4 \\ 1 \end{bmatrix} qx + \begin{bmatrix} 4 \\ 2 \end{bmatrix} q^3x^2 + \begin{bmatrix} 4 \\ 3 \end{bmatrix} q^6x^3 + q^{10}x^4. \end{aligned}$$

The polynomial $a_i(q)$ is the generating function for partitions into exactly i distinct parts, each of which is less than or equal to n.

Given a partition into exactly i distinct parts, each of which is less than or equal to n, we can subtract 1 from the smallest part, 2 from the second smallest part, and so on up to i from the largest part. This leaves us with an arbitrary partition into at most i parts, each of which is less than or equal to $n-i$, and we know that the generating function for these is $\begin{bmatrix} n-i+i \\ i \end{bmatrix} = \begin{bmatrix} n \\ i \end{bmatrix}$. We have subtracted $1+2+\cdots+i = i(i+1)/2$ from each partition, so the generating function for partitions into exactly i distinct parts, each of which is less than or equal to n, is

$$a_i(q) = q^{i(i+1)/2} \begin{bmatrix} n \\ i \end{bmatrix}.$$

Q.E.D.

We can use the q-binomial theorem to give a different proof of Jacobi's triple product identity. We take equation (3.5), set $n = 2m$,

$$(1+xq)(1+xq^2)\cdots(1+xq^{2m}) = \sum_{i=0}^{2m} \begin{bmatrix} 2m \\ i \end{bmatrix} q^{i(i+1)/2}\, x^i,$$

and then replace x by xq^{-m},

$$\begin{aligned}(1+xq^{1-m})&(1+xq^{2-m})\cdots(1+x)(1+xq)(1+xq^2)\cdots(1+xq^m)\\ &= \sum_{i=0}^{2m} \begin{bmatrix} 2m \\ i \end{bmatrix} q^{i(i+1)/2-mi}\, x^i.\end{aligned}$$

For $0 \le j \le m-1$, we rewrite $(1+xq^{-j})$ as $xq^{-j}(1+x^{-1}q^j)$:

$$\begin{aligned}x^m q^{-(0+1+2+\cdots+m-1)}&(1+x^{-1}q^{m-1})(1+x^{-1}q^{m-2})\cdots(1+x^{-1})\\ &\times (1+xq)(1+xq^2)\cdots(1+xq^m)\\ &= \sum_{i=0}^{2m} \begin{bmatrix} 2m \\ i \end{bmatrix} q^{i(i+1)/2-mi}\, x^i.\end{aligned}$$

We now move the powers of x and q to the other side by multiplying each side by $x^{-m}q^{m(m-1)/2}$,

$$\begin{aligned}(1+xq)(1+xq^2)\cdots(1+xq^m) &\times (1+x^{-1})(1+x^{-1}q)\cdots(1+x^{-1}q^{m-1})\\ = \sum_{i=0}^{2m} \begin{bmatrix} 2m \\ i \end{bmatrix}& q^{i(i+1)/2-mi+m(m-1)/2}\, x^{i-m}.\end{aligned}$$

We introduce a new variable of summation, $j = i - m$, and note that

$$j(j+1)/2 = (i-m)(i-m+1)/2 = i(i+1)/2 - mi + m(m-1)/2.$$

The q-binomial theorem implies that

$$\prod_{k=1}^{m}(1+xq^k)(1+x^{-1}q^{k-1}) = \sum_{j=-m}^{m} \begin{bmatrix} 2m \\ m+j \end{bmatrix} q^{j(j+1)/2}\, x^j. \tag{3.6}$$

This is a finite version of Jacobi's triple product identity.

We now let m go to infinity and require that $|q| < 1$. Since q^m approaches 0, the Gaussian polynomial becomes

$$\lim_{m\to\infty} \prod_{i=1}^{m+j} \frac{1-q^{m-j+i}}{1-q^i} = \prod_{i=1}^{\infty} \frac{1}{1-q^i}.$$

Exercises

3.1.1 Find a one-to-one correspondence between partitions into at most m parts, each less than or equal to n, and partitions into at most n parts, each less than or equal to m. This proves combinatorially that

$$\begin{bmatrix} m+n \\ m \end{bmatrix} = \begin{bmatrix} m+n \\ n \end{bmatrix}.$$

3.1.2 Verify equation (3.4).

3.1.3 This exercise and those up to and including exercise 3.1.10 outline Gauss's evaluation of the Gaussian sum $G(\alpha) = \sum \alpha^{j^2}$ where α is a primitive kth root of unity such as $e^{2\pi i/k}$ and the summation is taken over all residue classes, j, modulo k, where k is odd. This evaluation was Gauss's original reason for defining Gaussian polynomials.

Define

$$f(q,m) = 1 - \begin{bmatrix} m \\ 1 \end{bmatrix} + \begin{bmatrix} m \\ 2 \end{bmatrix} - \cdots + (-1)^m \begin{bmatrix} m \\ m \end{bmatrix}.$$

Use the recursive formula, $\begin{bmatrix} m \\ j \end{bmatrix} = \begin{bmatrix} m-1 \\ j \end{bmatrix} + q^{m-j}\begin{bmatrix} m-1 \\ j-1 \end{bmatrix}$, to prove that

$$f(q,m) = (1-q^{m-1})\, f(q,m-2).$$

Now prove by induction that

$$f(q,m)=\begin{cases}0, & \text{if } m \text{ is odd,}\\ (1-q)(1-q^3)\cdots(1-q^{m-1}), & \text{if } m \text{ is even.}\end{cases} \tag{3.7}$$

3.1.4 Let k be odd. Show that

$$\frac{1-\alpha^{k-j}}{1-\alpha^j}=-\alpha^{-j}.$$

Use this to prove that

$$\begin{bmatrix} k-1 \\ j \end{bmatrix}_\alpha = (-1)^j\alpha^{-j(j+1)/2},$$

and therefore

$$f(\alpha,k-1)=\sum_{j=0}^{k-1}\alpha^{-j(j+1)/2}.$$

3.1.5 Use the fact that if k is odd and α is a primitive kth root of unity, then so is α^{-2} to prove that

$$\begin{aligned} f(\alpha^{-2},k-1) &= \alpha^{-[(k+1)/2]^2}\sum_{j=0}^{k-1}\alpha^{[j+(k+1)/2]^2} \\ &= \alpha^{-[(k+1)/2]^2}G(\alpha). \end{aligned}$$

3.1.6 Use equation (3.7) to prove that

$$G(\alpha)=(\alpha-\alpha^{-1})(\alpha^3-\alpha^{-3})\cdots(\alpha^{k-2}-\alpha^{-(k-2)}). \tag{3.8}$$

3.1.7 Use the fact that $\alpha^{k-j}-\alpha^{-(k-j)}=-(\alpha^j-\alpha^{-j})$ to rewrite equation (3.8) as

$$G(\alpha)=(-1)^{(k-1)/2}(\alpha^2-\alpha^{-2})(\alpha^4-\alpha^{-4})\cdots(\alpha^{k-1}-\alpha^{-(k-1)}). \tag{3.9}$$

3.1.8 Combine equations (3.8) and (3.9) to show that

$$G(\alpha)^2=(-1)^{(k-1)/2}\alpha^{k(k-1)/2}\prod_{j=1}^{k-1}(1-\alpha^{-2j}). \tag{3.10}$$

Show that

$$\prod_{j=1}^{k-1}(x-\alpha^{-2j})=\frac{x^k-1}{x-1}=1+x+x^2+\cdots+x^{k-1},$$

and therefore

$$G(\alpha)^2 = (-1)^{(k-1)/2}k. \tag{3.11}$$

3.1.9 Use equation (3.9) to prove that

$$\begin{aligned} G(e^{2\pi i/k}) &= (-1)^{(k-1)/2} \prod_{j=1}^{(k-1)/2} (e^{4\pi ij/k} - e^{-4\pi ij/k}) \\ &= (-1)^{(k-1)/2}(2i)^{(k-1)/2} \prod_{j=1}^{(k-1)/2} \sin\frac{4\pi j}{k}. \end{aligned} \tag{3.12}$$

3.1.10 Prove that $\prod_{j=1}^{(k-1)/2} \sin(4\pi j/k)$ is positive for $k \equiv \pm 1 \pmod 8$ and negative for $k \equiv \pm 1 \pmod 8$. Combine this with equations (3.11) and (3.12) to prove that

$$G(e^{2\pi i/k}) = i^{[(k-1)/2]^2}\sqrt{k}. \tag{3.13}$$

3.1.11 Prove that a polynomial, $f(q)$, of degree m is palindromic if and only if $f(q) = q^m f(q^{-1})$. Use this characterization to prove that all Gaussian polynomials are palindromic.

3.1.12 Prove that in the Gaussian polynomial $\left[{m+n \atop m}\right]$, the coefficient of q^i, $0 \le i \le mn/2$, will always be the same as the coefficient of q^{mn-i} by finding a one-to-one correspondence between partitions of i into at most m parts, each of which is less than or equal to n, and partitions of $mn - i$ into at most m parts less than or equal to n.

3.1.13 The following identity for Gaussian polynomials was discovered by Kathy O'Hara (1990) and Doron Zeilberger (1989):

$$\begin{bmatrix} n+j \\ j \end{bmatrix} = \sum q^{m_1^2+\cdots m_j^2-j} \prod_{i=1}^{j} \begin{bmatrix} (n+2)i - M_{i-1} - M_{i+1} \\ m_i - m_{i+1} \end{bmatrix}, \tag{3.14}$$

where the sum is over all partitions of j: $m_1 + \cdots + m_j = j$, $m_1 \ge m_2 \ge \cdots \ge m_j \ge 0$, where $M_i = m_1 + \cdots + m_i$, and where we define $m_0 = M_0 = m_{j+1} = 0$. Prove that this equation is correct for $j = 1$ and $j = 2$. The next three exercises use this identity to prove that the Gaussian polynomial is unimodal.

3.1.14 We define the **mean power** of a polynomial to be the average of the highest and lowest powers with non-zero coefficients. The mean power of $3x^4 + x^5 - x^7 + 2x^9$ is $(4 + 9)/2 = 6.5$. Prove that the sum of two palindromic unimodal polynomials of mean power m is also a palindromic unimodal polynomial of mean power m.

3.1.15 Prove that the product of two palindromic unimodal polynomials of mean powers m_1 and m_2, respectively, is a palindromic unimodal polynomial of mean power $m_1 + m_2$.

3.1.16 Assume that for $b < j$ or $a < n + j$, $\begin{bmatrix} a \\ b \end{bmatrix}$ is a palindromic unimodal polynomial of mean power $b(a - b)/2$. Prove that each summand on the right side of equation (3.14) is a palindromic unimodal polynomial of mean power $nj/2$. Given equation (3.14), complete the proof that the Gaussian polynomial is unimodal.

3.1.17 Prove that equation (3.4) is correct whenever m is a positive integer. Show that n does not have to be a positive integer.

3.1.18 The notation $(q; q)_n$ is a special case of the more general

$$(a; q)_n = (1 - a)(1 - aq) \cdots (1 - aq^{n-1}).$$

With this notation, the q-binomial theorem, equation (3.5), can be written as

$$(xq; q)_n = \sum_{i=0}^{n} \begin{bmatrix} n \\ i \end{bmatrix} q^{i(i+1)/2}(-x)^i.$$

Explain why, if $|q| < 1$, then the q-binomial theorem is equivalent to

$$\frac{(x; q)_\infty}{(xq^n; q)_\infty} = \sum_{i=0}^{n} \begin{bmatrix} n \\ i \end{bmatrix} q^{i(i-1)/2}(-x)^i.$$

3.1.19 Interpret the coefficient of x^i in the power series expansion of

$$\frac{1}{(xq; q)_n} = \frac{1}{(1 - xq)(1 - xq^2) \cdots (1 - xq^n)}.$$

Use this interpretation to prove that

$$\frac{1}{(x; q)_n} = \sum_{i=0}^{\infty} \begin{bmatrix} n + i - 1 \\ i \end{bmatrix} x^i.$$

3.1.20 Prove that

$$(q^{n-i+1}; q)_i = (-1)^i q^{ni - i(i-1)/2} (q^{-n}; q)_i,$$

and then use this to prove that

$$\frac{(x; q)_\infty}{(xq^n; q)_\infty} = \sum_{i=0}^{\infty} \frac{(q^{-n}; q)_i}{(q; q)_i} x^i q^{ni}. \tag{3.15}$$

Why are we allowed to take the sum all the way to infinity?

3.1.21 In equation (3.15), restrict $|q| < 1$ and then show that when you take the limit of each side as n goes to infinity, you get

$$\prod_{k=1}^{\infty}(1 + xq^k) = \sum_{i=0}^{\infty} \frac{q^{i(i+1)/2}x^i}{(q;q)_i}. \tag{3.16}$$

3.1.22 Prove equation (3.16) by interpreting the coefficient of x^i in the power series expansion of $\prod_{k\geq 1}(1 + x\,q^k)$.

3.1.23 A result from complex analysis says that if two functions are both **analytic** (can be differentiated arbitrarily often) on the same open set and if they both agree on an infinite sequence of points that converges to a point inside this open set, then they must agree everywhere on this open set. Use this result and exercise 3.1.20 to prove that for any complex number a:

$$\frac{(ax;q)_\infty}{(x;q)_\infty} = \sum_{i=0}^{\infty} \frac{(a;q)_i}{(q;q)_i}\, x^i. \tag{3.17}$$

3.2 Inversion numbers

At the beginning of this chapter, we saw that there is a one-to-one correspondence between lattice paths from $(0,0)$ to (m,n) and sequences that consists of n 0s and m 1s. This implies that we have a correspondence between partitions into at most n parts each of which is less than or equal to m and such sequences. It is not particularly difficult to move back and forth between them. Each 0 represents a step to the north which marks off the end of one part in the partition. The size of that part is the number of 1s that lie to the left of that 0:

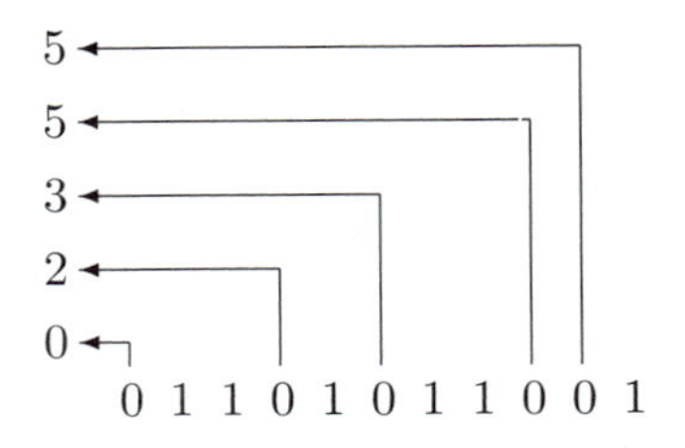

Each finite sequence of 0s and 1s corresponds to a partition. The sum of the parts in the partition is called the **inversion number** of the sequence, and we denote it by $\mathcal{I}$,

$$\mathcal{I}(01101011001) = 5 + 5 + 3 + 2 + 0 = 15.$$

As a corollary of Proposition 3.2, we have that

$$\sum_{s \in S(n,m)} q^{\mathcal{I}(s)} = \begin{bmatrix} m+n \\ m \end{bmatrix} = \prod_{i=1}^{m} \frac{1-q^{n+i}}{1-q^{i}}, \tag{3.18}$$

where $S(n,m)$ is the set of all sequences of n 0s and m 1s.

We can also define the inversion number of a sequence as the sum over all 1s in the sequence of the number of 0s after that 1. We note that this gives us the conjugate partition:

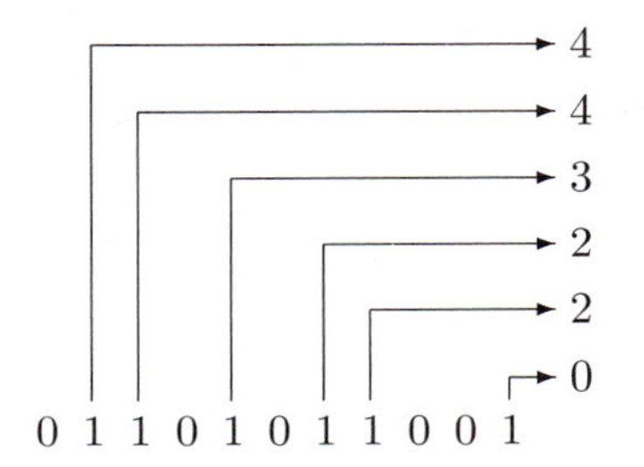

The q-multinomial coefficient

Given any sequence of integers, we take each term in the sequence and count the number of terms to its left that are strictly larger. The sum of these numbers is the **inversion number** of the sequence,

$$\mathcal{I}(362241536241) = 0+0+2+2+1+5+1+3+0+6+3+10 = 33.$$

The number of sequences with m_1 1s, m_2 2s, ..., m_n ns is given by the **multinomial coefficient**:

$$\binom{m_1+m_2+\cdots+m_n}{m_1, m_2, \ldots, m_n} = \frac{(m_1+m_2+\cdots+m_n)!}{m_1!\, m_2! \cdots m_n!}.$$

If instead of adding 1 for each sequence s, we add $q^{\mathcal{I}(s)}$, we get the **q-multinomial coefficient** defined by

$$\begin{bmatrix} m_1+m_2+\cdots+m_n \\ m_1, m_2, \ldots, m_n \end{bmatrix}_q = \sum_{s \in S} q^{\mathcal{I}(s)},$$

where $S = S(m_1, m_2, \ldots, m_n)$ is the set of all sequences with m_1 1s, m_2 2s, ..., m_n ns. As with the q-binomial coefficient, the subscript q is usually dropped unless something other than q is used as the base.

Proposition 3.4 *The q-multinomial coefficient is equal to the following rational product:*

$$\begin{bmatrix} m_1 + m_2 + \cdots + m_n \\ m_1, m_2, \ldots, m_n \end{bmatrix} = \frac{(q;q)_{m_1+m_2+\cdots+m_n}}{(q;q)_{m_1}(q;q)_{m_2}\cdots(q;q)_{m_n}} \tag{3.19}$$

Proof: We shall prove this proposition by induction on n, the number of distinct integers. The theorem is trivially true when $n = 1$, and it is Proposition 3.2 when $n = 2$. We assume that it is true up to and including $n-1$. Given a sequence s in $S(m_1, m_2, \ldots, m_n)$, s corresponds to a unique pair of sequences, $s' \in S(m_1, \ldots, m_{n-2}, m_{n-1} + m_n)$ and $t \in S(m_{n-1}, m_n)$ where s' is obtained from s by relabeling each n as an $n-1$ and t is the subsequence of s that consists of the $n-1$s and the ns. We see that t tells us which $n-1$s in s' were originally ns.

The inversion number of s is the inversion number of s' plus the inversion number of t. We therefore have that

$$\sum_{s \in S} q^{\mathcal{I}(s)} = \sum_{s' \in S'} q^{\mathcal{I}(s')} \sum_{t \in T} q^{\mathcal{I}(t)}.$$

We can use our induction hypothesis:

$$\sum_{s \in S} q^{\mathcal{I}(s)} = \frac{(q;q)_{m_1+m_2+\cdots+(m_{n-1}+m_n)}}{(q;q)_{m_1}(q;q)_{m_2}\cdots(q;q)_{m_{n-1}+m_n}} \cdot \frac{(q;q)_{m_{n-1}+m_n}}{(q;q)_{m_{n-1}}(q;q)_{m_n}}.$$

Q.E.D.

The inversion number of a permutation

We can represent a permutation of $\{1, 2, \ldots, n\}$ as a sequence in which the ith entry is the image of i. For example, the permutation

$$\sigma(1) = 1, \quad \sigma(2) = 3, \quad \sigma(3) = 5, \quad \sigma(4) = 2, \quad \sigma(5) = 4$$

can be written as the sequence 13524. We can then define the **inversion number of the permutation** as the inversion number of the corresponding sequence in which $m_1 = m_2 = \cdots m_n = 1$. For example,

$$\mathcal{I}(13524) = 0 + 0 + 0 + 2 + 1 = 3.$$

Corollary 3.5 *If we let $\mathcal{S}_n$ denote the set of permutations on n letters, then*

$$\sum_{\sigma \in \mathcal{S}_n} q^{\mathcal{I}(\sigma)} = \frac{(1-q)(1-q^2)\cdots(1-q^n)}{(1-q)^n}. \tag{3.20}$$

It is worth noting that if we take the limit as q approaches 1 of each side of this equation, the right side becomes $n!$. This polynomial is therefore referred to as the **q-factorial**, $[n!]_q$.

$$\begin{aligned}
[1!]_q &= 1, \\
[2!]_q &= 1+q, \\
[3!]_q &= 1+2q+2q^2+q^3, \\
[4!]_q &= 1+3q+5q^2+6q^3+5q^4+3q^5+q^6, \\
[5!]_q &= 1+4q+9q^2+15q^3+20q^4+22q^5+20q^6+15q^7 \\
&\quad +9q^8+4q^9+q^{10}.
\end{aligned}$$

The Gaussian polynomial can be expressed as a ratio of q-factorials:

$$\begin{bmatrix} m+n \\ m \end{bmatrix} = \frac{[(m+n)!]_q}{[m!]_q\,[n!]_q}.$$

The inversion number of a permutation is a measure of how far the permutation is from the identity. Given any permutation written as a sequence of integers, it is always possible to turn it into the identity permutation by transposing adjacent elements. For example, three transpositions will turn 13524 back into 12345:

$$13524 \longrightarrow 13254 \longrightarrow 12354 \longrightarrow 12345.$$

While there are other routes back to the identity, for example,

$$13524 \longrightarrow 13254 \longrightarrow 13245 \longrightarrow 12345,$$

all of them require at least three transpositions. The inversion number of a permutation is also the smallest number of transpositions of adjacent elements required to turn the permutation back into the identity.

There is an important observation that we shall use in the next section: any transposition of two adjacent elements in a permutation will either increase the inversion number by 1 or decrease it by 1. From this it follows that a transposition of any two elements in a permutation will change the parity of the inversion number.

The inversion number of a permutation appears in the calculation of determinants. The determinant of the matrix

$$(a_{ij})_{i,j=1}^{n}$$

can be written as

$$\det\,(a_{ij})_{i,j=1}^{n} = \sum_{\sigma\in\mathcal{S}_n} (-1)^{\mathcal{I}(\sigma)} \prod_{i=1}^{n} a_{i\,\sigma(i)}.$$

For our purposes, it is convenient to take this as the definition of the determinant.

The inversion number of a matrix

If we describe a permutation by the permutation matrix of 0s and 1s, then there is an easy way of calculating the inversion number directly from this matrix. The inversion number will be the number of pairs of 1s in this matrix for which one of the 1s lies to the right and above the other. There are three such pairs in the matrix

$$\begin{pmatrix} 1 & 0 & 0 & 0 & 0 \\ 0 & 0 & 1 & 0 & 0 \\ 0 & 0 & 0 & 0 & 1 \\ 0 & 1 & 0 & 0 & 0 \\ 0 & 0 & 0 & 1 & 0 \end{pmatrix}.$$

Equivalently, we could calculate the inversion number by taking all pairs of matrix entries for which one of them lies to the right and above the other (in a 5×5 matrix, there are 100 such pairs; see exercise 3.2.16), multiplying each pair of entries together (which gives us 0 unless they are both 1), and then adding up all of these products.

This is a procedure that can be applied to any matrix. In particular, we can use it to define the **inversion number of an alternating sign matrix**. The inversion number of

$$\begin{pmatrix} 0 & 1 & 0 & 0 & 0 \\ 0 & 0 & 1 & 0 & 0 \\ 1 & -1 & 0 & 0 & 1 \\ 0 & 1 & -1 & 1 & 0 \\ 0 & 0 & 1 & 0 & 0 \end{pmatrix}.$$

is 5. There are seven pairs whose product is $+1$ and two pairs whose product is -1.

No one has found a nice identity for

$$\sum_{B \in \mathcal{A}_n} q^{\mathcal{I}(B)},$$

where $\mathcal{A}_n$ is the set of $n \times n$ alternating sign matrices, but if we let $N(B)$ be the number of -1s in B, then

$$\sum_{B \in \mathcal{A}_n} q^{\mathcal{I}(B)}(1 + q^{-1})^{N(B)}$$

can be factored into a product of linear polynomials. In exercise 3.2.14 you are asked to use *Mathematica* to experiment with this sum and make your own conjecture about the formula. This sum will appear again in a more general context in Section 3.5.

Exercises

3.2.1 Given a sequence, s, of 0s and 1s, we define the **major index** of s, $\mathrm{maj}(s)$, to be the sum of the positions of the 1s that are immediately followed by 0s. For example, $\mathrm{maj}(01101011001) = 3 + 5 + 8 = 16$. Find $\mathrm{maj}(11001010100)$. What is the inversion number of each of these sequences?

3.2.2 The *Mathematica* commands given below will calculate `Inv`, the inversion number, `Maj`, the major index, and `Perms[m,n]`, a list of all permutations of m 0s and n 1s. The arguments of `Inv` and `Maj` must be entered as a list such as `Maj[{0,1,1,0,1,0,1}]`. The command `Apply[Maj,Perms[2,3]]` will calculate the major index for each of the permutations of two 0s and three 1s. Compare the sequence of answers with what you get when you apply `Inv` to `Perms[2,3]`. Compare the sequences of values that you get with other numbers of 0s and 1s.

```
Perms[x_,y_]:=Permutations[Join[Table[0,{x}],
   Table[1,{y}]]];
Inv[x_]:=Count[Flatten[Map[Drop[x,#]&,
   Flatten[Position[x,1]]]],0];
Maj[x_]:=Plus@@Intersection[Flatten[Position[x,1]],
   Flatten[Position[x,0]]-1]
```

3.2.3 Prove that the following gives a bijection between the set of partitions into at most n parts, each less than or equal to m, and the set of sequences of m 0s and n 1s in such a way that the sum of the parts in the partition is equal to the major index of the corresponding sequence. Show that this implies that

$$\sum_{s \in S(m,n)} q^{\mathrm{maj}(s)} = \begin{bmatrix} m+n \\ m \end{bmatrix}.$$

Let λ be a partition with at most n parts, each less than or equal to m. Let $\lambda_1 \geq \lambda_2 \geq \cdots \geq \lambda_n \geq 0$ be the parts. Start with a sequence of m 0s. For $j = 1, \ldots, n$: If $\lambda_j \geq j$, insert a 1 into the sequence so that there are $\lambda_j - j$ 0s and no 1s to the left; if $\lambda_j < j$, insert a 1 into the sequence so that there are λ_j 1s to the right and no 0 immediately to the right of the new 1.

3.2.4 Define the major index of any sequence to be the sum of the positions of letters that are immediately followed by a smaller letter. For example $\text{maj}(324142231) = 1+3+5+8 = 17$. Prove that

$$\sum_{s \in S(m_1,\ldots,m_k)} q^{\text{maj}(s)} = \begin{bmatrix} m_1 + m_2 + \cdots + m_k \\ m_1,\ m_2,\ \ldots,\ m_k \end{bmatrix}.$$

3.2.5 Prove that if $n \geq 2$, then

$$\sum_{\sigma \in \mathcal{S}_n} (-1)^{\mathcal{I}(\sigma)} = 0.$$

3.2.6 Use induction to prove that the inversion number counts the least number of transpositions of adjacent elements required to turn the given permutation back into the identity.

3.2.7 Differentiate each side of equation (3.20) to prove that

$$\sum_{\sigma \in \mathcal{S}_n} \mathcal{I}(\sigma) = n!\, n(n-1)/4.$$

Then find a direct proof of this fact that does not use the generating function.

3.2.8 Given an alternating sign matrix, B, let r_i be the set of integers in the ith row, counted from the top, of a monotone triangle that corresponds to B. Let $r_i - r_{i+1}$ denote the set of elements of r_i that are not in r_{i+1}. Let $N(B)$ be the number of -1s in the alternating sign matrix B. Prove that

$$N(B) = \sum_{i=1}^{n-1} |r_i - r_{i+1}|.$$

3.2.9 Let r_i be defined as in exercise 3.2.8. Let $[[k]] = \{1, 2, \ldots, k\}$ and $[[0]] = \{\}$. Define

$$\begin{aligned} P(r_i, r_{i+1}) &= \sum_{k \in r_{i+1} - r_i} |[[k-1]] - r_{i+1}|, \\ M(r_i, r_{i+1}) &= \sum_{k \in r_i - r_{i+1}} |[[k-1]] - r_{i+1}|. \end{aligned}$$

Prove that the inversion number of B is equal to

$$\sum_{i=1}^{n-1} P(r_i, r_{i+1}) - M(r_i, r_{i+1}).$$

3.2.10 Define

$$MT_n(x,q) = \sum_{B \in \mathcal{A}_n} q^{\mathcal{I}(B)} x^{N(B)}.$$

Using the results of exercises 3.2.8 and 3.2.9, prove that

$$\begin{aligned} MT_n(x,q) &= \sum_{j=1}^{n} MT_n(\{j\};x,q), \quad \text{where} \\ MT_n(r_i;x,q) &= \sum{}^{*} q^{P(r_i,r_{i+1})-M(r_i,r_{i+1})} x^{|r_i - r_{i+1}|} \\ &\qquad \times MT_n(r_{i+1};x,q), \end{aligned}$$

and $MT_n(\{1,2,\ldots,n\};x,q) = 1$. We define $\sum^*$ to be the sum over all possible rows r_{i+1} that could lie directly below r_i.

3.2.11 The recursion given in exercise 3.2.10 can be encoded in the following *Mathematica* program. Use it to see if there is any pattern to `ASMf[n,1,q]`, the sum over all $n \times n$ alternating sign matrices of q to the inversion number.

```
nextlist[n_,alist_]:=Select[Distribute[Apply[Range,
   Partition[Prepend[Append[alist,n],1], 2, 1], 1],
   List],# == Union[#] &];
Pl[alist_,blist_]:=
   Plus@@Map[Length[Complement[Range[#-1],blist]]&,
   Complement[blist,alist]];
Mi[alist_,blist_]:=
   Plus@@Map[Length[Complement[Range[#-1],blist]]&,
   Complement[alist,blist]];
Nmo[alist_,blist_]:=Length[Complement[alist,blist]];
ASMf[n_,x_,q_]:=
   Expand[Sum[q^(j-1) MTf[{j},n,x,q],{j,n}]];
MTf[l_,n_,x_,q_]:=If[l==Range[n],1,Plus@@Map[
   q^(Pl[l,#]-Mi[l,#]) x^Nmo[l,#] MTf[#,n,x,q]&,
   nextlist[n,l]] ]
```

3.2.12 Choose small values of k and look for any patterns in `ASMf[n,k,1]` where k is a fixed integer. What can you prove?

3.2.13 Prove that $\mathcal{I}(B) \geq 0$ for any alternating sign matrix B.

3.2.14 Find a closed formula for `ASMf[n,1 + 1/q,q]`,

$$\sum_{B \in \mathcal{A}_n} q^{\mathcal{I}(B)} (1+q^{-1})^{N(B)}.$$

3.2.15 Prove that $\mathcal{I}(B) \geq N(B)$ for any alternating sign matrix B.

Figure 3.1. Xavier Viennot (*left*) and Ira Gessel.

3.2.16 Show that in an $n \times n$ matrix, there are exactly $\binom{n}{2}^2$ pairs of entries in which one is north and east of the other.

3.3 Plane partitions

We now have the tools to prove Theorem 1.1, the product form of the generating function for plane partitions in $\mathcal{B}(r,s,t)$. This will be done in two stages. First, we use our lattice paths to prove that the generating function for plane partitions that fit inside $\mathcal{B}(r,s,t)$ is given by the determinant

$$\det\left(q^{i(i-j)}\begin{bmatrix} s+t \\ s-i+j \end{bmatrix}\right)_{i,j=1}^{r}.$$

Then we use Krattenthaler's formula to prove that this determinant can be expressed as the product

$$\prod_{(i,j,k)\in\mathcal{B}(r,s,t)} \frac{1-q^{i+j+k-1}}{1-q^{i+j+k-2}}.$$

It is left as an exercise to show that as r, s, and t approach infinity, this product becomes $\prod 1/(1-q^j)^j,\ j \geq 1$.

MacMahon's proof of the generating function for plane partitions in $\mathcal{B}(r,s,t)$ was published in 1912. It has seen many improvements and clarifications. In the early 1980s, Xavier Viennot and Ira Gessel showed how to simplify the proof by presenting the arguments in terms of lattice paths (Gessel and Viennot 1985).

Their work was presaged by Bernt Lindström (1973), who worked with lattice paths in representation theory, but who missed their potential as a combinatorial tool. Gessel and Viennot had begun their work independently. They joined efforts when they discovered that they had each found the same key results.

We begin with MacMahon's representation of a plane partition which can be thought of as a bird's-eye view of the stacks of unit cubes. We replace each stack by a single integer that counts the number of cubes in that particular stack. The plane partition given in Figure 1.3 on page 10 is represented by

$$\begin{array}{cccccc} 6 & 5 & 5 & 4 & 3 & 3 \\ 6 & 4 & 3 & 3 & 1 & \\ 6 & 4 & 3 & 1 & 1 & \\ 4 & 2 & 2 & 1 & & \\ 3 & 1 & 1 & & & \\ 1 & 1 & 1. & & & \end{array}$$

It is this particular two-dimensional or planar representation that leads to the name *plane partition.* The fact that the cubes are stacked into a corner is reflected in the requirement that the entries are weakly decreasing as we move to the right across any row or down any column.

If our plane partition fits inside the box $\mathcal{B}(r, s, t)$, then this planar representation has at most r rows, at most s columns, and the largest part is less than or equal to t. Each row is, itself, an ordinary partition into at most s parts, each less than or equal to t. Allowing for some of the partitions to be empty, we see that each plane partition inside $\mathcal{B}(r, s, t)$ is a sequence of r ordinary partitions, each of which has at most s parts, and each of these parts is less than or equal to t.

If our r partitions could be chosen independently, then the generating function would be

$$\begin{bmatrix} s+t \\ s \end{bmatrix}^r .$$

Of course, they are not independent. The jth part of the ith partition must be less than or equal to the jth part in the $i-1$st partition. We can account for this relationship between successive partitions if we represent each partition as a lattice path, and then *nest* the successive lattice paths.

The first path (corresponding to the partition given by the top line of the plane partition) goes from $(0, 0)$ to (t, s). The next path above it (corresponding to the second line) goes from $(-1, 1)$ to $(t-1, s+1)$. The condition that the j part of the second partition must be less than or equal to the jth part of the first partition is equivalent to the condition that these two lattice paths are not permitted to touch. The third line of the plane partition is encoded as a path from $(-2, 2)$ to $(t-2, s+2)$ which is not allowed to touch any other path.

We count our lattice paths beginning with the topmost path, the one that corresponds to the lowest partition. In general, the ith lattice path goes from $(-r+i, r-i)$ to $(t-r+i, s+r-i)$ and corresponds to the $(r+1-i)$th line of the plane partition. If we consider our sample plane partition as sitting inside the box $\mathcal{B}(7, 6, 6)$ (seven lines with at most six parts to each line and each part less than or equal to 6), then it is represented as seven lattice paths shown in Figure 3.2. The first of these lattice paths corresponds to the empty partition which is always represented by a lattice path that heads due north and then due east. We shall call any such collection of lattice paths a nest of lattice paths, or, more simply, a **nest**.

How to count non-intersecting lattice paths

Rather than trying to establish the generating function, we shall begin with the slightly simpler problem of counting the total number of ways we can construct r nested, non-intersecting lattice paths, each of which goes t steps to the east and s to the north. If our paths were allowed to intersect, then it would be easy to count them. The number would be

$$\binom{s+t}{s}^r.$$

To get rid of the paths that do intersect, we have to do an inclusion–exclusion argument of a rather special type, a sum over permutations.

We consider a larger set of nests, each of which contains r lattice paths which always move east or north; each path begins at one of the vertices $(-r+j, r-j)$, $1 \leq j \leq r$; each path ends at one of the vertices $(t-r+i, s+r-i)$, $1 \leq i \leq r$; and no two paths start at the same vertex or end at the same vertex. This allows for some very tangled nests, such as the one shown in Figure 3.3.

Each nest defines a permutation of the integers 1 through r as follows: The ith path is the path that ends at $(t-r+i, s+r-i)$. The jth starting

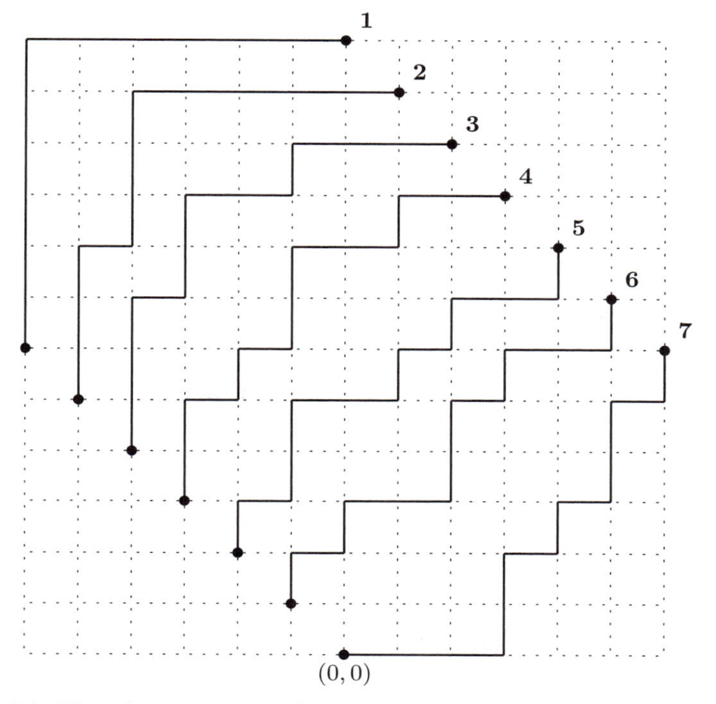

Figure 3.2. The plane partition of 75 represented as a nest of seven lattice paths.

point is $(-r+j, r-j)$. If the ith path starts at the jth starting point, then we define $\sigma(i) = j$. The tangled nest given in Figure 3.3 corresponds to the permutation

$$3517246.$$

If we are given the permutation, σ, then the ith lattice path in the nest must begin at vertex $(-r + \sigma(i), r - \sigma(i))$ and end at vertex $(t - r + i, s + r - i)$. It takes $t + i - \sigma(i)$ steps to the east and $s - i + \sigma(i)$ steps to the north. The number of such lattice paths is

$$\binom{t+s}{s-i+\sigma(i)}.$$

The number of nests that correspond to the permutation σ is

$$\prod_{i=1}^{l} \binom{t+s}{s-i+\sigma(i)}.$$

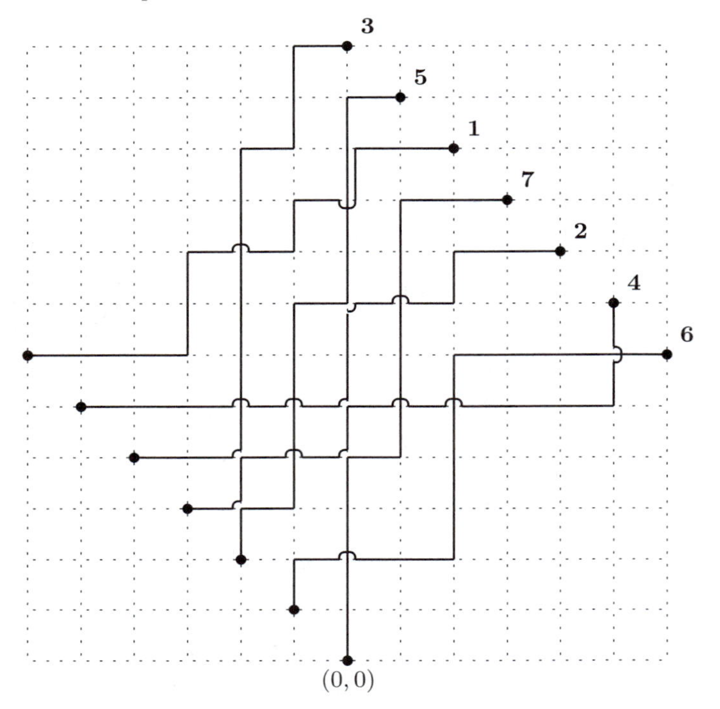

Figure 3.3. A tangled nest of seven lattice paths.

Theorem 3.6 *Given positive integers r, s, and t, the total number of plane partitions that fit inside $\mathcal{B}(r,s,t)$ is*

$$\sum_{\sigma \in \mathcal{S}_r} (-1)^{\mathcal{I}(\sigma)} \prod_{i=1}^{r} \binom{t+s}{s-i+\sigma(i)} = \det\left(\binom{t+s}{s-i+j}\right)_{i,j=1}^{r}. \qquad (3.21)$$

Proof: Our determinant is the sum over all possible nests where each nest is counted as $+1$ if the inversion number of σ is even and as -1 if the inversion number of σ is odd. Any nest of non-intersecting lattice paths must correspond to the identity permutation, and so is counted as $+1$. We must show that everything else in this summation cancels. We shall do this by separating all nests of intersecting lattice paths into pairs of nests which cancel each other out; one nest in each pair will have an even inversion number and other will have an odd inversion number.

Given a nest with at least one point where two paths meet, we choose the intersection point that is furthest to the right. If there is more than one intersection point in this column, we choose the one that is highest.

For the example given in Figure 3.3, this is the point at $(5, 6)$. We now switch the paths at this point (if they cross, we change them so that they just meet; if they just meet, we change them so that they cross):

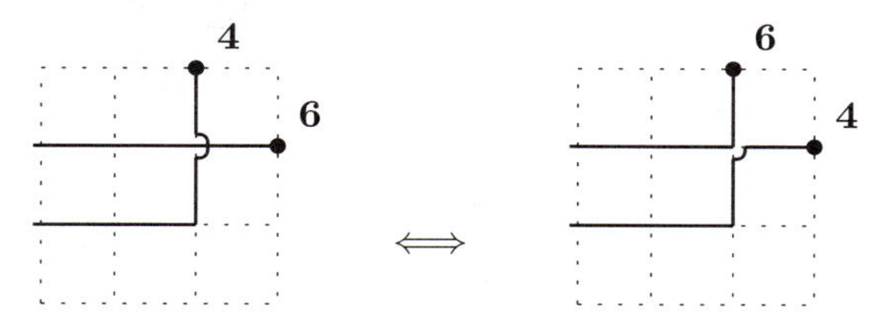

The effect of this switch is to transpose two adjacent terms in the corresponding permutation. In our example, the new permutation is

$$3517264.$$

Q.E.D.

The generating function

To turn this result into a generating function, we have to do more than just replace each binomial coefficient with a Gaussian polynomial. A problem arises when we switch the tails of two intersecting lattice paths. If we assign to each nest not just the number 1, but a power of q that corresponds to the number being partitioned, then changing those tails will change the number that is partitioned. Taking our example (see Fig. 3.4), the first path corresponds to the partition $8+8+3+1$ (the closed circles). The second path corresponds to $3+3+3+3$ (the open circles). All of these parts add up to 32. After switching the tails of the paths the first partition becomes $8+3+1$, and the second partition becomes $6+3+3+3+3$. These add to 30.

In Figure 3.4, we can see what is happening. Because of the way we chose our intersection point, the two paths that we switch will always be consecutive (see exercise 3.3.2); let us call them paths i and $i+1$. If $\sigma(i) > \sigma(i+1)$, then when we switch the tails, we pick up an additional $\sigma(i) - \sigma(i+1)$. If, on the other hand, $\sigma(i) < \sigma(i+1)$, then switching the tails decreases the total partition by $\sigma(i+1) - \sigma(i)$. In either case, switching tails increases the total amount partitioned by $\sigma(i) - \sigma(i+1)$.

We can compensate for this if we define the total amount that is being

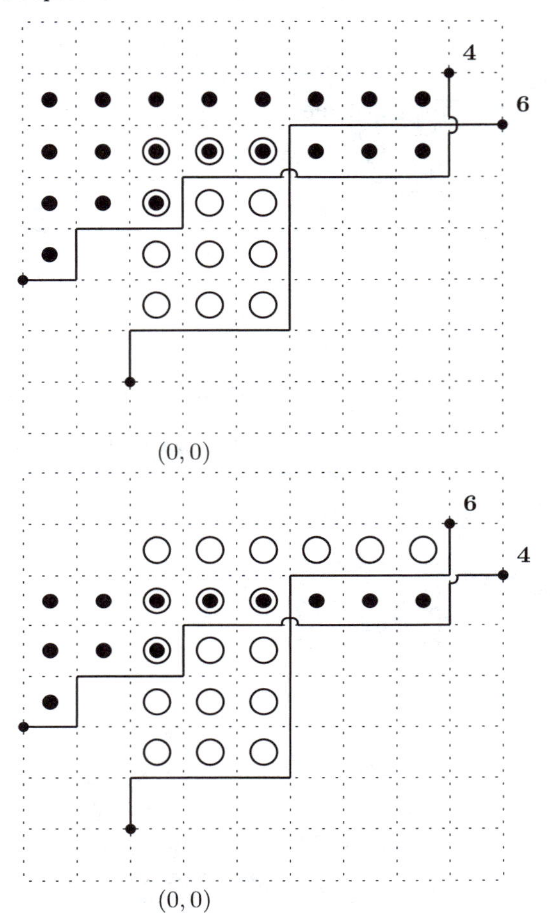

Figure 3.4. Switching the tails of the last two lattice paths.

partitioned as the sum of the amounts partitioned by each lattice path *plus*

$$\sum_{i=1}^{r} i(i-\sigma(i)).$$

When we transpose the ith and $(i+1)$st integers in the permutation, this summation *decreases* by $\sigma(i)-\sigma(i+1)$. When we have a nest of non-intersecting paths, σ is the identity permutation and this summation is zero. For a nest of non-intersecting paths, the amount partitioned is the sum of the amounts partitioned by each lattice path.

We have found a determinant that gives us the generating function we seek.

Theorem 3.7 *Given positive integers* r, s, *and* t, *the generating function for plane partitions that fit inside* $\mathcal{B}(r,s,t)$ *is*

$$\sum_{\sigma \in \mathcal{S}_r} (-1)^{\mathcal{I}(\sigma)} \prod_{i=1}^{r} q^{i(i-\sigma(i))} \begin{bmatrix} t+s \\ s-i+\sigma(i) \end{bmatrix} = \det \left(q^{i(i-j)} \begin{bmatrix} t+s \\ s-i+j \end{bmatrix} \right)_{i,j=1}^{r}. \tag{3.22}$$

Conclusion

To evaluate this determinant, we use Krattenthaler's formula (Theorem 2.9 on page 67). For each i, we factor $(q;q)_{s+t}/(q;q)_{s-i+r}$ out of the ith row of our matrix, and for each j we factor $1/(q;q)_{t+j-1}$ out of the jth column. The determinant that we want to evaluate is equal to

$$\prod_{i=1}^{r} \frac{(q;q)_{s+t}}{(q;q)_{s-i+r}(q;q)_{t+i-1}} \times \det \Big(q^{i(i-j)}(1-q^{s-i+r})(1-q^{s-i+r-1})\cdots(1-q^{s-i+j+1}) \cdot (1-q^{t+i-j+1})(1-q^{t+i-j+2})\cdots(1-q^{t+i-1}) \Big)_{i,j=1}^{r}.$$

We now isolate the q^i in each of the binomials inside the determinant so that this becomes

$$\begin{aligned} &\prod_{i=1}^{r} \frac{(q;q)_{s+t}}{(q;q)_{s-i+r}(q;q)_{t+i-1}} \\ &\quad \times \det \Big((-1)^{j-1} q^{i(i-j)} q^{-i(r-j)} q^{(j-1)(2t-j)/2} \\ &\qquad \cdot (q^i - q^{s+r})(q^i - q^{s+r-1})\cdots(q^i - q^{s+j+1}) \\ &\qquad \cdot (q^i - q^{-t+j-1})(q^i - q^{-t+j-2})\cdots(q^i - q^{-t+1}) \Big)_{i,j=1}^{r} \\ &= (-1)^{r(r-1)/2} q^{\sum_{i=1}^{r} t(i-1)-ri+(i^2+i)/2} \prod_{i=1}^{r} \frac{(q;q)_{s+t}}{(q;q)_{s-i+r}(q;q)_{t+i-1}} \\ &\quad \times \det \Big((q^i - q^{s+r})(q^i - q^{s+r-1})\cdots(q^i - q^{s+j+1}) \\ &\qquad \cdot (q^i - q^{-t+j-1})(q^i - q^{-t+j-2})\cdots(q^i - q^{-t+1}) \Big)_{i,j=1}^{r}. \end{aligned}$$

We can now apply Krattenthaler's formula with $x_i = q^i$, $a_j = -q^{s+j}$, and $b_j = -q^{-t+j-1}$. The proof of Theorem 1.3 has been reduced to verifying that

$$(-1)^{r(r-1)/2} q^{\sum_{i=1}^{r} t(i-1)-ri+(i^2+i)/2} \prod_{i=1}^{r} \frac{(q;q)_{s+t}}{(q;q)_{s-i+r}(q;q)_{t+i-1}} \times \prod_{1\le i<j\le r} (q^i - q^j) \prod_{2\le i\le j\le r} (-q^{-t+i-1} + q^{s+j}) = \prod_{i=1}^{r}\prod_{j=1}^{t} \frac{1-q^{i+j+s-1}}{1-q^{i+j-1}}. \tag{3.23}$$

I leave this as an exercise.

Q.E.D.

This has been our first major proof. As promised, it is a hybrid of a lattice path argument to reduce the problem to evaluating a determinant, and then a specialization of a result proved by symmetry arguments to evaluate the determinant. In broad terms, this will be the structure of each of our three big proofs, the proofs of Conjectures 4, 6, and 1.

Exercises

3.3.1 Prove that

$$\lim_{r,s,t\to\infty} \prod_{(i,j,k)\in\mathcal{B}(r,s,t)} \frac{1-q^{i+j+k-1}}{1-q^{i+j+k-2}} = \prod_{j=1}^{\infty} \frac{1}{(1-q^j)^j}.$$

3.3.2 Prove that the chosen intersection point will always be an intersection of two consecutive paths.

3.3.3 Prove that

$$f(\sigma) = \sum_{i=1}^{l} i(i-\sigma(i))$$

is the unique function of $\sigma \in \mathcal{S}_l$ that is zero when σ is the identity permutation and decreases by $\sigma(i) - \sigma(i+1)$ when we transpose the ith and $i+1$st integers in the permutation.

3.3.4 Prove that

$$(-1)^{r(r-1)/2} q^{\sum_{i=1}^{r} t(i-1)-ri+(i^2+i)/2} \times \prod_{1\le i<j\le r} (q^i - q^j) \prod_{2\le i\le j\le r} (-q^{-t+i-1} + q^{s+j})$$

$$= \prod_{1 \le i < j \le r} (1 - q^{j-i}) \prod_{2 \le i \le j \le r} (1 - q^{s+t+1+j-i}).$$

3.3.5 Prove that

$$\prod_{i=1}^{r} \frac{1}{(q;q)_{t+i-1}} \prod_{1 \le i < j \le r} (1 - q^{j-i}) = \prod_{i=1}^{r} \prod_{j=1}^{t} \frac{1}{1 - q^{i+j-1}}.$$

3.3.6 Complete the proof of equation (3.23) by showing that

$$\prod_{i=1}^{r} \frac{(q;q)_{s+t}}{(q;q)_{s-i+r}} \prod_{2 \le i \le j \le r} (1 - q^{s+t+1+j-i}) = \prod_{i=1}^{r} \prod_{j=1}^{t} (1 - q^{s+i+j-1}).$$

3.4 Cyclically symmetric plane partitions

In this section we shall lay the groundwork for proving Conjectures 6, 7, and 9. We shall derive the generating functions, expressed as determinants, for cyclically symmetric plane partitions, for descending plane partitions, and for each of these when we specify the number of times that the largest possible part appears. We shall see how Mills, Robbins, Rumsey evaluated these determinants when we get to Section 5.3.

We have seen (Section 1.3) how to separate a cyclically symmetric plane partition into shells and then encode it by describing the bottom level of each shell. If the plane partition fits inside an $r \times r \times r$ box, then the bottom level of the outermost shell must consist of at most r parts, each less than or equal to r, with the additional restriction that the largest part must equal the number of parts on this level. This restriction guarantees that three copies of the bottom level can be put together to form the outermost shell. From Figure 1.7 on page 20, we see that on the next level, formed by taking the bottom of the next shell, there is no part directly above the largest part of the bottom level, and each part on the second level must be strictly less than the part directly below.

If we encode each level as an ordinary partition, then each successive level is indented one from the previous level, and we must have strict decrease down each column. Our example, the cyclically symmetric plane partition of 75, encodes as

$$\begin{array}{cccccc} 6 & 5 & 5 & 4 & 3 & 3 \\ & 3 & 2 & 2 & & \\ & & 1 & & & \end{array}.$$

A plane partition in which each row is indented one from the row above is called a **shifted plane partition**. It is a **strict shifted plane partition** if we have strict decrease as we move down any column. From this decomposition into shells, we see that the number of cyclically symmetric plane partitions that fit inside $\mathcal{B}(r,r,r)$ is equal to the number of strict shifted plane partitions with largest part less than or equal to r and such that the number of parts in each row is equal to the largest part in that row.

We now represent each of these rows as a lattice path, but the situation is somewhat different from the last section because the number of parts – and thus the ending point – for each lattice path depends on the size of the largest part in that row. We need to know the largest part in each row before we can determine how to set up our lattice paths. Let m be the number of rows in the strict shifted plane partition, and let a_i, $1 \leq i \leq m$, be the largest part (and also the number of parts) in the ith row, counted from the bottom up so that $1 \leq a_1 < a_2 < \cdots < a_m \leq r$. We call a_i the ith **row leader**.

Since we know the largest part in each row, we can remove it and concentrate on the partition that remains. In the ith row from the bottom, this will be a partition with exactly $a_i - 1$ parts. We can subtract one from each of these parts. This leaves us with an arbitrary partition into at most $a_i - 1$ parts, each of which is less than or equal to $a_i - 1$. We know that the number of such partitions is

$$\binom{2a_i - 2}{a_i - 1}.$$

The strict shifted plane partition with row leaders $1 \leq a_1 < a_2 < \cdots < a_m \leq r$ is represented by m lattice paths. The ith lattice path encodes the partition in the ith line. It is a partition into at most $a_i - 1$ parts, each less than or equal to $a_i - 1$.

Specifically, the ith lattice path goes from $(0, r - a_i)$ to $(a_i - 1, r - 1)$. In our example, the row leaders are $a_1 = 1$, $a_2 = 3$, $a_3 = r = 6$. The partitions to be encoded as lattice paths are the empty partition, $1 + 1$, and $4 + 4 + 3 + 2 + 2$; see Figure 3.5.

To count the number of strict shifted plane partitions with row leaders $a_1, a_2, \ldots, a_m$, we count all nests of lattice paths with starting points at $(0, r - a_j)$ and ending points at $(a_i - 1, r - 1)$, so that no two lattice paths have the same start or end points. As in the last section, each nest of lattice paths determines a permutation of $\{1, 2, \ldots, m\}$. If we take the number of nests of lattice paths for which the inversion number is even and subtract the number of nests of lattice paths for which the inversion

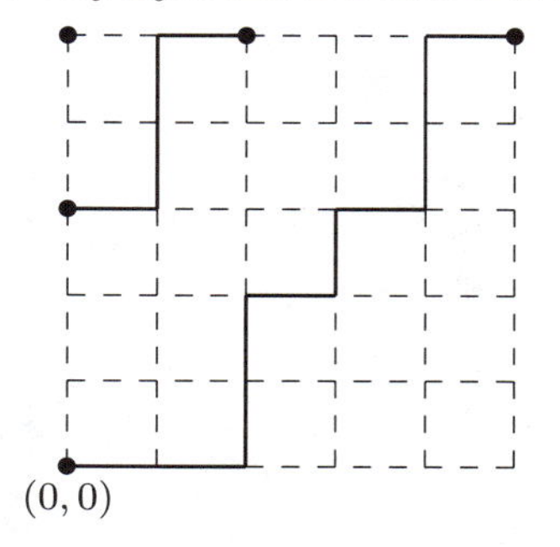

Figure 3.5. The cyclically symmetric plane partition of 75 encoded as a nest of strict shifted plane partitions.

number is odd, we are left with the number of nests of non-intersecting lattice paths. Given a nest of lattice paths with at least one intersection point, we find the other nest in exactly the same manner as before: We find the intersection points that are furthest to the right, and among all intersection points in the same column we take the highest. We then interchange the tails of these two paths. This gives us the following result.

Lemma 3.8 *The number of strict shifted plane partitions with parts less than or equal to r, with m rows, and such that the ith row, $1 \leq i \leq m$, starts with the part a_i and has exactly a_i parts is*

$$\sum_{\sigma \in \mathcal{S}_m} (-1)^{\mathcal{I}(\sigma)} \prod_{i=1}^{m} \binom{a_i + a_{\sigma(i)} - 2}{a_i - 1} = \det\left(\binom{a_i + a_j - 2}{a_i - 1} \right)_{i,j=1}^{m}.$$

We now translate this into a generating function for cyclically symmetric plane partitions. Because all paths start with the same x-coordinate, switching tails does not change the sum of the parts, and so we do not need the extra factor of $q^{i(i-j)}$ in the entries of the determinant. The row leaders of the strict shifted plane partition correspond to the **skeleton** of the shells, the darkened blocks in Figure 3.6. A row leader of size a_i represents $3a_i - 2$ cubes in the plane partition. The remaining partition in the ith row, a partition into at most $a_i - 1$ parts less than or equal to $a_i - 1$, determines how the remainder of that shell is filled in. This partition is reproduced three times, once on each side of the shell, so we need to replace q by q^3 in the generating function for partitions into at most $a_i - 1$ parts, each less than or equal to $a_i - 1$. The generating

Figure 3.6. Darkened blocks show the skeleton of the shells of the cyclically symmetric plane partition.

function for all possible cyclically symmetric shells with row leader a_i is

$$q^{3a_i-2}\begin{bmatrix}2a_i-2\\a_i-1\end{bmatrix}_{q^3}.$$

The generating function for all nests of lattice paths with row leaders $a_1, a_2, \ldots, a_m$ and for which the ith path starts at $(0, r-a_{\sigma(i)})$ is given by

$$\prod_{i=1}^{m} q^{3a_i-2}\begin{bmatrix}a_i+a_{\sigma(i)}-2\\a_i-1\end{bmatrix}_{q^3}.$$

Lemma 3.9 *The generating function for cyclically symmetric plane partition that fit inside* $\mathcal{B}(r,r,r)$*, with* m *shells and row leaders* $1 \le a_1 < a_2 < \cdots < a_m \le r$ *is*

$$\sum_{\sigma\in\mathcal{S}_m}(-1)^{\mathcal{I}(\sigma)}\prod_{i=1}^{m} q^{3a_i-2}\begin{bmatrix}a_i+a_{\sigma(i)}-2\\a_i-1\end{bmatrix}_{q^3}$$

$$= \det\left(q^{3a_i-2}\begin{bmatrix}a_i+a_j-2\\a_i-1\end{bmatrix}_{q^3}\right)_{i,j=1}^{m}.$$

Summing over sets of row leaders

To get the generating function for cyclically symmetric plane partitions, we sum the generating function given in Lemma 3.9 over all possible sets of rows. In other words, we sum over all subsets of $\{1,2,\ldots,r\}$:

$$\sum_{\{a_1,a_2,\ldots,a_m\}\subseteq\{1,2,\ldots,r\}} \sum_{\sigma\in\mathcal{S}_m} (-1)^{\mathcal{I}(\sigma)} \prod_{i=1}^{m} q^{3a_i-2} \begin{bmatrix} a_i + a_{\sigma(i)} - 2 \\ a_i - 1 \end{bmatrix}_{q^3}.$$

The next theorem shows how this can be accomplished by taking the determinant of the matrix that is constructed by first putting

$$q^{3i-2} \begin{bmatrix} i+j-2 \\ i-1 \end{bmatrix}_{q^3}$$

in row i, column j, and then adding 1 to each entry on the main diagonal.

Theorem 3.10 *Let G_r denote the matrix*

$$G_r = \left(q^{3i-2} \begin{bmatrix} i+j-2 \\ i-1 \end{bmatrix}_{q^3} \right)_{i,j=1}^{r}. \tag{3.24}$$

The generating function for cyclically symmetric plane partitions that fit inside $\mathcal{B}(r,r,r)$ is given by

$$\det(I_r + G_r) = \sum_{\sigma\in\mathcal{S}_r} (-1)^{\mathcal{I}(\sigma)} \prod_{i=1}^{r} \left(\delta_{i\sigma(i)} + q^{3i-2} \begin{bmatrix} i+\sigma(i)-2 \\ i-1 \end{bmatrix}_{q^3} \right), \tag{3.25}$$

where $\delta_{ij} = 1$ if $i = j$, $= 0$ if $i \neq j$. If we place the additional restriction that there are exactly k parts of size r, $0 \le k \le r$, then the generating function is the determinant of $G_{k,r}$, the matrix formed by replacing the last row of $I_r + G_r$ by $(0,\ldots,0,1)$ when $k = 0$ and by

$$q^{3k(r-1)+1} \left(\begin{bmatrix} r-1-k \\ 1-k \end{bmatrix}_{q^3}, \begin{bmatrix} r-k \\ 2-k \end{bmatrix}_{q^3}, \ldots, \begin{bmatrix} r-2-k \\ r-k \end{bmatrix}_{q^3} \right)$$

when $0 < k \le r$.

Proof: If we expand the summation side of the generating function in equation (3.25), we get a sum over all permutations on r letters and, for each i such that $\sigma(i) = i$, a choice of either the first term, 1, or the second term,

$$q^{3i-2} \begin{bmatrix} 2i-2 \\ i-1 \end{bmatrix}_{q^3}.$$

If $\sigma(i)$ is not equal to i or if $\sigma(i) = i$ and we choose the second term (the one that involves the Gaussian polynomial), then i is an element of our subset $\{a_1, a_2, \ldots, a_m\}$. If $\sigma(i) = i$ and we choose the first term, 1, then i is not an element of our subset. This gives us our sum over all possible subsets, including the full set $\{1, 2, \ldots, r\}$ when we never choose 1, and the empty set when σ is the identity permutation and we choose 1 every time.

Given $\{a_1, a_2, \ldots, a_m\}$ and a permutation, σ, on m letters, σ is embedded in a permutation, τ, on all r letters. The permutation τ is defined by

1. if $j \in \{a_1, a_2, \ldots, a_m\}$, say $j = a_i$, then $\tau(j) = a_{\sigma(i)}$,
2. if $j \notin \{a_1, a_2, \ldots, a_m\}$, then $\tau(j) = j$.

To complete our proof, we need to show that $(-1)^{\mathcal{I}(\sigma)} = (-1)^{\mathcal{I}(\tau)}$. This is equivalent to proving that the inversion numbers of σ and τ have the same parity.

As an example, let $r = 10$ and take $\{2, 5, 6, 8, 10\}$ as the subset with $m = 5$. The permutation $\sigma = 51423 \in \mathcal{S}_5$ has inversion number 6. The elements of our set are ordered as

$$10\ 2\ 8\ 5\ 6.$$

This also has 6 inversions. To find the corresponding permutation in $\mathcal{S}_{10}$, we first place the missing elements,

$$1\ \_\ 3\ 4\ \_\ \_\ 7\ \_\ 9\ \_,$$

and then insert the elements from our set in the specified order

$$\tau \ = \ 1\ 10\ 3\ 4\ 2\ 8\ 7\ 5\ 9\ 6.$$

This has inversion number 16.

Each transposition of adjacent elements in σ corresponds to a transposition of two elements that might not be adjacent in τ, but *any* transposition changes the parity of the inversion number (see exercise 3.4.1). We can restore τ to the identity using $\mathcal{I}(\sigma)$ transpositions, each of which changes $\mathcal{I}(\tau)$ by an odd amount. It follows that $\mathcal{I}(\sigma)$ and $\mathcal{I}(\tau)$ have the same parity.

This completes the proof of the first part of Theorem 3.10, the generating function for cyclically symmetric plane partitions that fit inside $\mathcal{B}(r, r, r)$. We now consider the second part of the theorem where we specify that there are k parts of size r.

The key to evaluating our determinant and so proving Conjecture 6 is to work with the generating function in which we specify the number of times r appears in the representation as a strict shifted plane partition. This number, which we shall denote by k, could be as small as 0 or as large as r.

If k is zero: In this case our strict shifted plane partition has no parts of size r. In other words, r is not an element of the set $\{a_1, \ldots, a_m\}$. Our generating function corresponds to those terms in the expansion of the determinant for which $\sigma(r) = r$ and we have chosen the first term, 1. We let G_r^* denote G_r with the bottom row replaced by a row of 0s. The generating function when $k = 0$ is the determinant of $I_r + G_r^*$.

If k is positive: In this case our strict shifted plane partition has at least one part of size r. This means that r is in our set $\{a_1, \ldots, a_m\}$ and so $a_m = r$. The nest of m lattice paths will have the mth path go from $(0, r - a_{\sigma(m)})$ to $(r-1, r-1)$. More than that, we know that this path ends with a step to the east followed by $k-1$ steps to the north. In other words, the last path must pass through the point $(r-2, r-k)$, and once it reaches that point, the rest of its tail is uniquely determined.

The generating function for paths from $(0, r - a_{\sigma(m)})$ to $(r-2, r-k)$ is

$$\begin{bmatrix} r - 2 + a_{\sigma(m)} - k \\ a_{\sigma(m)} - k \end{bmatrix}.$$

The mth path then takes one step to the east followed by $k-1$ steps north, which contributes $q^{(r-1)(k-1)}$ to the generating function,

$$q^{(r-1)(k-1)} \begin{bmatrix} r - 2 + a_{\sigma(m)} - k \\ a_{\sigma(m)} - k \end{bmatrix}.$$

The generating function for paths from $(0, r - a_{\sigma(m)})$ to $(a_m - 1, r-1)$ occupies the a_mth row and $a_{\sigma(m)}$th column of our matrix. Therefore, the entry in the rth row and the j column of our matrix must be

$$q^{3r-2}\, q^{3(r-1)(k-1)} \begin{bmatrix} r - 2 + j - k \\ j - k \end{bmatrix}_{q^3} = q^{3k(r-1)+1} \begin{bmatrix} r - 2 + j - k \\ j - k \end{bmatrix}_{q^3}.$$

Note that if $a_{\sigma(m)} = j$ is less than k, then there are no possible nests of lattice paths.

Q.E.D.

The generating function for descending plane partitions

A similar argument will also enable us to represent the generating function for descending plane partitions that fit inside an $r \times r \times r$ box as a determinant.

Theorem 3.11 *Let H_r denote the matrix*

$$H_r = \left(q^{i+1} \begin{bmatrix} i+j \\ j-1 \end{bmatrix} \right)_{i,j=1}^{r}. \tag{3.26}$$

The generating function for descending plane partitions that fit inside $\mathcal{B}(r,r,r)$ is given by

$$\det(I_{r-1} + H_{r-1}) = \sum_{\sigma \in \mathcal{S}_{r-1}} (-1)^{\mathcal{I}(\sigma)} \prod_{i=1}^{r-1} \left(\delta_{i\sigma(i)} + q^{i+1} \begin{bmatrix} i+\sigma(i) \\ i+1 \end{bmatrix} \right),$$

where $\delta_{ij} = 1$ if $i = j$, $= 0$ if $i \neq j$. If we place the additional restriction that there are exactly k parts of size r, $0 \leq k \leq r-1$, then the generating function is the determinant of $H_{k,r-1}$, the matrix formed by replacing the last row of $I_{r-1} + H_{r-1}$ by $(0, \ldots, 0, 1)$ when $k = 0$ and by

$$q^{kr} \left(\begin{bmatrix} r-k \\ 1-k \end{bmatrix}, \begin{bmatrix} r+1-k \\ 2-k \end{bmatrix}, \ldots, \begin{bmatrix} 2r-2-k \\ r-1-k \end{bmatrix} \right)$$

when $0 < k \leq r-1$.

The proof of this theorem parallels that of Theorem 3.10 and most of it is left for the exercises. We note that a_1, the smallest row leader, must be at least 2. Since we have only $r-1$ possible choices for row leaders, our generating function will be the determinant of an $(r-1) \times (r-1)$ matrix, and the row leaders are chosen from $\{i+1 \mid 1 \leq i \leq r-1\}$. We also note that if a_i is the row leader for the ith row, then after we remove this part, this row consists of at most $a_i - 2$ parts, each of which is less than or equal to a_i, and such partitions are represented by lattice paths from $(0, r-a_i)$ to $(a_i, r-2)$.

The restriction that the ith row must have at least a_{i-1} parts is equivalent to the condition that no two paths intersect at a point on the line $x = 0$. This is subsumed under the more general condition that no two paths are allowed to intersect.

The tough part of finding these generating functions lies in evaluating the determinants. What makes the determinants difficult to evaluate is the 1 added to each term on the main diagonal. Without it, we get determinants that are easily evaluated using Krattenthaler's formula.

Exercises

3.4.1 Prove that any transposition of two elements in a permutation can be executed by performing an odd number of adjacent transpositions.

3.4.2 Let f be any function defined on all pairs of integers taken from $\{1, 2, \ldots, n\}$. Prove that

$$\det\left(\delta_{ij} + f(a_i, a_j)\right) = \sum_{\{a_1, a_2, \ldots, a_m\} \subseteq \{1, 2, \ldots, r\}} \sum_{\sigma \in \mathcal{S}_m} (-1)^{\mathcal{I}(\sigma)} \prod_{i=1}^{m} f(a_i, a_{\sigma(i)}).$$

3.4.3 In exercise 1.2.8 of Section 1.2, you were asked to show that the conjectured generating function for cyclically symmetric plane partitions that fit inside $\mathcal{B}(r, r, r)$ can be written as

$$\prod_{i=1}^{r} \frac{1 - q^{2i+t-1}}{1 - q^{2i-1}} \prod_{1 \leq i < j \leq r} \frac{1 - q^{2i+2j+2t-2}}{1 - q^{2i+2j-2}}.$$

Use *Mathematica* to show that this conjectured generating function agrees with $\det(I_r + G_r)$ for $1 \leq r \leq 5$.

3.4.4 Andrews's conjecture, Conjecture 7, is that the generating function for descending plane partitions is given by

$$\prod_{1 \leq i \leq j \leq r} \frac{1 - q^{r+i+j-1}}{1 - q^{2i+j-1}}.$$

Use *Mathematica* to prove that for $2 \leq r \leq 6$, this is equal to the determinant of $I_{r-1} + H_{r-1}$.

3.4.5 Let A_1, A_2, ..., A_t be $n \times n$ matrices that are identical except for the last row which is $(a_{n1}^{(k)}, a_{n2}^{(k)}, \ldots, a_{nn}^{(k)})$ in A_k. Prove that the sum of the determinants of the A_k is equal to the determinant of the matrix that agrees with these matrices in the first $n-1$ rows and whose nth row is the vector sum

$$\sum_{k=1}^{t} (a_{n1}^{(k)}, a_{n2}^{(k)}, \ldots, a_{nn}^{(k)}).$$

3.4.6 Prove that

$$\sum_{k=1}^{r} \binom{r-2+j-k}{j-k} = \binom{r-2+j}{j-1}, \quad 1 \leq j \leq r. \tag{3.27}$$

Hint: Use the combinatorial description of $\binom{r-2+j}{j-1}$ as the number of paths from $(0, 0)$ to $(r-1, j-1)$.

3.4.7 Prove that

$$\sum_{k=1}^{r} q^{k(r-1)} \begin{bmatrix} r-2+j-k \\ j-k \end{bmatrix} = q^{r-1} \begin{bmatrix} r-2+j \\ j-1 \end{bmatrix}, \quad 1 \le j \le r. \tag{3.28}$$

Use this result and exercise 3.4.5 to prove that

$$\sum_{k=0}^{r} \det(G_{k,r}) = \det(I_r + G_r).$$

3.4.8 Prove that

$$\sum_{k=0}^{r} \det(H_{k,r}) = \det(I_r + H_r).$$

3.4.9 Show that there is a one-to-one correspondence between descending plane partitions with row leaders $2 \le a_1 < a_2 < \cdots < a_m \le r$ and nests of non-intersecting lattice paths in which the ith path goes from $(0, r - a_i)$ to $(a_i, r - 2)$.

3.4.10 Prove that the generating function for descending plane partitions with row leaders $2 \le a_1 < a_2 < \cdots < a_m \le r$ is given by

$$\sum_{\sigma \in \mathcal{S}_m} (-1)^{\mathcal{I}(\sigma)} \prod_{i=1}^{m} q^{a_i} \begin{bmatrix} a_i + a_{\sigma(i)} - 2 \\ a_i \end{bmatrix}$$

which is equal to $\det\left(q^{a_i} \begin{bmatrix} a_i+a_j-2 \\ a_i \end{bmatrix}\right)$.

3.4.11 Use the result in exercise 3.4.10 to prove that the generating function for descending plane partitions that fit inside $\mathcal{B}(r,r,r)$ is the determinant of $I_{r-1} + H_{r-1}$.

3.4.12 Explain why the generating function for descending plane partitions in $\mathcal{B}(r,r,r)$ with no parts of size r is equal to the determinant of $H_{0,r-1}$.

3.4.13 Let $1 \le k \le r-1$. Prove that the generating function for descending plane partitions with row leaders $2 \le a_1 < a_2 < \cdots < a_m = r$ and exactly k parts of size r is given by

$$\sum_{\sigma \in \mathcal{S}_m} (-1)^{\mathcal{I}(\sigma)} q^{kr} \begin{bmatrix} r + a_{\sigma(i)} - k - 2 \\ a_{\sigma(m)} - k - 1 \end{bmatrix} \prod_{i=1}^{m-1} q^{a_i} \begin{bmatrix} a_i + a_{\sigma(i)} - 2 \\ a_i \end{bmatrix}.$$

3.4.14 Complete the proof that the generating function for descending plane partitions that fit inside $\mathcal{B}(r,r,r)$ and have exactly k parts of size r is the determinant of $H_{k,r-1}$.

3.5 Dodgson's algorithm

David Robbins and Howard Rumsey discovered alternating sign matrices in the course of analyzing and generalizing a determinant evaluation algorithm published by Charles Lutwidge Dodgson (1866). Dodgson's algorithm is based on the Desnanot–Jacobi adjoint matrix theorem.

Given a matrix, M, we let M_j^i denote the matrix that remains when the ith row and jth column of M are deleted. If we wish to delete more than one row or column, the numbers of the deleted rows are listed as superscripts, the numbers of the deleted columns as subscripts. We shall use absolute values around a matrix to denote its determinant: $|M|$ means $\det(M)$.

Theorem 3.12 (Desnanot–Jacobi adjoint matrix theorem) *If M is an $n \times n$ matrix, then*

$$|M| \; |M_{1\,n}^{1\,n}| = |M_1^1| \; |M_n^n| - |M_n^1| \; |M_1^n| \, . \tag{3.29}$$

Lagrange discovered this theorem for $n = 3$ in 1773. Desnanot proved it for $n \leq 6$ in 1819. Jacobi published the general theorem in 1833. Jacobi played an important role in the development of the theory of determinants. In 1841, he published three major papers on the subject: "On the formation and properties of determinants" (1841a) which contains a second proof of Theorem 3.12, "On functional determinants" (1841b) in which the Jacobian matrix – a matrix of partial derivatives – is defined, and "On alternating functions and their division by the product of the differences of the elements" (1841c) which includes the Jacobi–Trudi identity that we shall see in Chapter 4. According to Muir (1906), "The three memoirs together constitute an excellent treatise on the subject [of determinants], and are known to have been markedly influential in spreading a knowledge of it among mathematicians."

Jacobi, now thirty-seven years old, was reaching his zenith. Two years later, he became gravely ill with diabetes. His friend and colleague, Peter Gustav Lejeune Dirichlet, arranged for a special gift from the kaiser, Friedrich Wilhelm IV, so that Jacobi could go to Italy for his health. Dirichlet and Jacobi spent a year there. When they returned, Dirichlet arranged for Jacobi to join him at the University of Berlin. In 1851, Jacobi contracted smallpox and died.

This section begins with a proof of the Desnanot–Jacobi theorem, followed by an explanation of Dodgson's algorithm for calculating determinants. We shall then see how this algorithm leads to the λ-determinant of Robbins and Rumsey, and conclude with a remarkable theorem that ties all of this to alternating sign matrices.

Proof of the adjoint matrix theorem

The proof of Theorem 3.12 uses the **cofactor matrix**. Given an $n \times n$ matrix, M, the cofactor matrix, M^C, is the matrix whose entry in the ith row and jth column is, up to sign, the determinant of the matrix that remains when the jth row and ith column of M are removed:

$$M^C = \begin{pmatrix} |M_1^1| & -|M_1^2| & \cdots & (-1)^{n+1}|M_1^n| \\ -|M_2^1| & |M_2^2| & \cdots & (-1)^{n+2}|M_2^n| \\ \vdots & \vdots & & \vdots \\ (-1)^{n+1}|M_n^1| & (-1)^{n+2}|M_n^2| & \cdots & |M_n^n| \end{pmatrix}.$$

If we take the dot product of the ith row of M with the ith column of the cofactor matrix, we get

$$(-1)^{i+1}\left(m_{i1}|M_1^i| - m_{i2}|M_2^i| + \cdots + (-1)^{n-1}m_{in}|M_n^i|\right) = |M|.$$

On the other hand, if we take the dot product of the jth row of M with the ith column of M^C, $j \neq i$, this gives us the determinant of the matrix in which row i has been replaced by a second copy of row j, and the determinant of this matrix is zero:

$$m_{j1}|M_1^i| - m_{j2}|M_2^i| + \cdots + (-1)^{n-1}m_{jn}|M_n^i| = 0, \qquad (j \neq i).$$

If we multiply a matrix by its cofactor matrix, we get

$$M \cdot M^C = \begin{pmatrix} |M| & 0 & 0 & \cdots & 0 \\ 0 & |M| & 0 & \cdots & 0 \\ \vdots & & & \ddots & \vdots \\ 0 & 0 & 0 & \cdots & |M| \end{pmatrix}.$$

Since the determinant of a product of two matrices is the product of the two determinants, we see that $|M| \cdot |M^C| = |M|^n$. This equation can be viewed as a polynomial identity in n^2 variables, $m_{1,1}$ through $m_{n,n}$, which means that we can divide each side by the non-zero polynomial represented by $|M|$:

$$|M^C| = |M|^{n-1}. \tag{3.30}$$

We prove Theorem 3.12 by repeating what we have just done with a modified form of the cofactor matrix. Starting with M^C, we replace the entry in row i, column j $(2 \le j \le n-1)$ by 0 if $i \neq j$ and by 1 if $i = j$. We call this matrix $M^\star$:

$$M^\star = \begin{pmatrix} |M_1^1| & 0 & 0 & \cdots & 0 & (-1)^{n+1}|M_1^n| \\ -|M_2^1| & 1 & 0 & \cdots & 0 & (-1)^{n+2}|M_2^n| \\ |M_3^1| & 0 & 1 & \cdots & 0 & (-1)^{n+3}|M_3^n| \\ \vdots & \vdots & \vdots & \ddots & \vdots & \vdots \\ (-1)^n|M_{n-1}^1| & 0 & 0 & \cdots & 1 & -|M_{n-1}^n| \\ (-1)^{n+1}|M_n^1| & 0 & 0 & \cdots & 0 & |M_n^n| \end{pmatrix}.$$

If we multiply M by $M^\star$, we get

$$M \cdot M^\star = \begin{pmatrix} |M| & m_{12} & m_{13} & \cdots & 0 \\ 0 & m_{22} & m_{23} & \cdots & 0 \\ 0 & m_{32} & m_{33} & \cdots & 0 \\ \vdots & \vdots & \vdots & \ddots & \vdots \\ 0 & m_{n2} & m_{m3} & \cdots & |M| \end{pmatrix}.$$

The determinant of this product is $|M|^2\,|M_{1\,n}^{1\,n}|$, and the determinant of $M^\star$ is $|M_1^1|\,|M_n^n| - |M_n^1|\,|M_1^n|$. We have demonstrated that

$$|M|\left(|M_1^1|\,|M_n^n| - |M_n^1|\,|M_1^n|\right) = |M|^2\,|M_{1\,n}^{1\,n}|.$$

Again, we view this as a polynomial identity in $m_{1,1}$ through $m_{n,n}$ and divide each side by $|M|$.

Q.E.D.

Dodgson's determinant evaluation algorithm

The Reverend Charles L. Dodgson, better known under his pen name of Lewis Carroll, used Theorem 3.12 to devise an algorithm for calculating determinants that required calculating only 2×2 determinants. He called his method **condensation of determinants**.

We are given an $n \times n$ matrix, $M = (a_{i,j})$, whose determinant we are to find. The algorithm works on pairs of matrices, (A, B). Initially, A is our matrix M and B is an $n-1 \times n-1$ matrix with each entry equal to 1.

If none of the entries of B is zero, then we construct a new pair of matrices: A' which is $n-1 \times n-1$ and B' which is $n-2 \times n-2$, as follows:

$$\begin{aligned} a'_{i,j} &= (a_{i,j}a_{i+1,j+1} - a_{i,j+1}a_{i+1,j})/b_{i,j}, \quad i,j = 1,\ldots,n-1 \\ b'_{i,j} &= a_{i+1,j+1}, \quad i,j = 1,\ldots,n-2 \end{aligned}$$

In other words, A' consists of the determinants of the 2×2 adjacent submatrices of A divided by the corresponding element of B, and B' is A with the outer edge of entries removed.

This procedure is iterated until either

- at least one of the entries in B is zero (we are stuck at this point; Dodgson suggests that we go back to the original matrix, reorder the rows, and try again)
- or, A has a single entry and B is empty. In this case, the value in A is the determinant.

Example:

$$A = \begin{pmatrix} 1 & -2 & -1 & 3 \\ 2 & 1 & -1 & 2 \\ -1 & -2 & 1 & -3 \\ 0 & -1 & -1 & 2 \end{pmatrix} \qquad B = \begin{pmatrix} 1 & 1 & 1 \\ 1 & 1 & 1 \\ 1 & 1 & 1 \end{pmatrix},$$

$$\Downarrow$$

$$A' = \begin{pmatrix} 5 & 3 & 1 \\ -3 & -1 & 1 \\ 1 & 3 & -1 \end{pmatrix} \qquad B' = \begin{pmatrix} 1 & -1 \\ -2 & 1 \end{pmatrix},$$

$$\Downarrow$$

$$A'' = \begin{pmatrix} 4 & -4 \\ 4 & -2 \end{pmatrix} \qquad B'' = \begin{pmatrix} -1 \end{pmatrix},$$

$$\Downarrow$$

$$A''' = \begin{pmatrix} -8 \end{pmatrix} \qquad B''' = \begin{pmatrix} \quad \end{pmatrix}.$$

The determinant is -8.

Why it works

A **minor** of a matrix is a determinant of a submatrix obtained by deleting entire rows and columns. An **adjacent minor** is the determinant of a submatrix in which the rows and columns that remain were consecutive in the original matrix. The **central minor** is the determinant of the submatrix obtained by removing the first and last rows and columns, what we have denoted by $M_{1,n}^{1,n}$. When we say that a minor is $j \times j$, we mean that it is the determinant of a $j \times j$ submatrix. We know that A' consists of the 2×2 adjacent minors of M. If we want to find the 3×3 adjacent minors of A, Theorem 3.12 tells us that we can use the 2×2 adjacent minors, as long as we then divide by the central element:

$$\det \begin{pmatrix} 1 & -2 & -1 \\ 2 & 1 & -1 \\ -1 & -2 & 1 \end{pmatrix}$$

$$= \frac{\begin{vmatrix} 1 & -2 \\ 2 & 1 \end{vmatrix}\begin{vmatrix} 1 & -1 \\ -2 & 1 \end{vmatrix} - \begin{vmatrix} -2 & -1 \\ 1 & -1 \end{vmatrix}\begin{vmatrix} 2 & 1 \\ -1 & -2 \end{vmatrix}}{1}$$

$$= \frac{5(-1) - 3(-3)}{1} = 4.$$

When we calculate the 2×2 adjacent minors of A' and divide each one by the corresponding central element of the 3×3 adjacent submatrix of A, we are actually calculating the 3×3 adjacent minors of A. The matrix A'' is the matrix of 3×3 adjacent minors of A.

In general, we let $A^{(1)} = A'$, $A^{(2)} = A''$, ..., and suppose that $A^{(k-1)}$ is the matrix whose entries are the $k \times k$ adjacent minors of M and that $B^{(k-1)}$ is the central submatrix of the matrix of $k-1 \times k-1$ adjacent minors. Theorem 3.12 implies that $A^{(k)}$ is the matrix whose entries are the $k+1 \times k+1$ adjacent minors of M. By definition, $B^{(k)}$ is the central submatrix of the matrix of $k \times k$ adjacent minors. By induction, this is true for all k. In particular, $A^{(n-1)}$ is the matrix of $n \times n$ adjacent minors of M. But there is only one $n \times n$ minor of M. It is the determinant of M.

The λ-determinant

The central feature of Dodgson's algorithm is that it builds arbitrary determinants out of 2×2 determinants. Robbins and Rumsey asked the question, What happens if we generalize the definition of a 2×2 determinant to

$$\begin{vmatrix} a_{11} & a_{12} \\ a_{21} & a_{22} \end{vmatrix}_{\lambda} = a_{11}a_{22} + \lambda a_{12}a_{21}?$$

Using the idea of Dodgson's algorithm, we can use induction to build the λ-determinant for an arbitrary square matrix.

As an example, the 3×3 λ-determinant is

$$\begin{aligned}
&\begin{vmatrix} a_{11} & a_{12} & a_{13} \\ a_{21} & a_{22} & a_{23} \\ a_{31} & a_{32} & a_{33} \end{vmatrix}_\lambda \\
&= \left[(a_{11}\,a_{22} + \lambda a_{12}\,a_{21})(a_{22}\,a_{33} + \lambda a_{23}\,a_{32}) \right. \\
&\qquad \left. + \lambda(a_{12}\,a_{23} + \lambda a_{13}\,a_{22})(a_{21}\,a_{32} + \lambda a_{22}\,a_{31})\right] / a_{22} \\
&= a_{11}\,a_{22}\,a_{33} + \lambda a_{12}\,a_{21}\,a_{33} + \lambda a_{11}\,a_{23}\,a_{32} + \lambda^2 a_{12}\,a_{21}\,a_{23}\,a_{32}/a_{22} \\
&\qquad + \lambda a_{12}\,a_{23}\,a_{21}\,a_{32}/a_{22} + \lambda^2 a_{13}\,a_{21}\,a_{32} + \lambda^2 a_{12}\,a_{23}\,a_{31} \\
&\qquad + \lambda^3 a_{13}\,a_{22}\,a_{31}.
\end{aligned}$$

This occurred during the early 1980s. David Robbins had just received his first computer algebra package (ALTRAN) and decided to play with it, letting it find the next several λ-determinants. He stared at the output for several days before realizing that there was a pattern. These λ-determinants could all be expressed as a sum of monomials in the a_{ij} and their inverses, each monomial multiplied by a polynomial in λ. The monomials that appeared were quite restricted. They were all of the form

$$\prod_{i,j=1}^{n} a_{ij}^{B_{ij}},$$

where B_{ij} was the entry of row i, column j of an $n \times n$ matrix of 1s, -1s, and 0s with row sums 1, column sums 1, and alternating signs on non-zero entries. In other words, (B_{ij}) was an alternating sign matrix. Furthermore, the polynomials in λ had a simple expression in terms of the inversion number of B and the number of -1s in B. He and Howard Rumsey eventually proved the following theorem (1986).

Theorem 3.13 *Let M be an $n \times n$ matrix with entries a_{ij}, $\mathcal{A}_n$ the set of $n \times n$ alternating sign matrices, $\mathcal{I}(B)$ the inversion number of B, and $N(B)$ the number of -1s in B, then*

$$|M|_\lambda = \sum_{B \in \mathcal{A}_n} \lambda^{\mathcal{I}(B)} (1 + \lambda^{-1})^{N(B)} \prod_{i,j=1}^{n} a_{ij}^{B_{ij}}. \tag{3.31}$$

Out of curiosity, Robbins now asked himself how many summands there would be. The alternating sign matrix conjecture was about to be discovered.

The generalized Vandermonde product

In exercise 3.5.6, you are asked to use the inductive definition of the λ-determinant to prove that

$$\left|x_j^{n-i}\right|_\lambda = \prod_{1\le i<j\le n} (x_i + \lambda x_j).$$

When we combine this with Theorem 3.13, we see that

$$\prod_{1\le i<j\le n} (x_i + \lambda x_j) = \sum_{B\in\mathcal{A}_n} \lambda^{\mathcal{I}(B)}(1+\lambda^{-1})^{N(B)} \prod_{i,j=1}^{n} x_j^{(n-i)B_{ij}}. \quad (3.32)$$

Soichi Okada (1993) has found similar formulæ generalizing the other Weyl denominator formulæ. Okada's results for the root systems $\mathbf{B_n}$ and $\mathbf{C_n}$ involve sums over alternating sign matrices that are invariant under 180° rotation.

Exercises

3.5.1 Use Dodgson's algorithm to evaluate the determinant of

$$\begin{pmatrix} 2 & 0 & 1 & 3 \\ -1 & 2 & 1 & -2 \\ 0 & -1 & 1 & 3 \\ 2 & 4 & -3 & 2 \end{pmatrix}.$$

3.5.2 Prove that for $1 \le i < r \le n$ and $1 \le j < s \le n$, the Desnanot–Jacobi adjoint matrix theorem can be generalized to

$$|M| = \frac{|M_j^i|\,|M_s^r| - |M_s^i|\,|M_j^r|}{|M_{j,s}^{i,r}|}.$$

3.5.3 The theorem that Jacobi proved in 1833 was actually more general than Theorem 3.12. Let $(M^C)^{m+1,\ldots,n}_{m+1,\ldots,n}$ be the upper left $m \times m$ corner of the cofactor matrix. Jacobi proved that

$$\left|(M^C)^{m+1,\ldots,n}_{m+1,\ldots,n}\right| = |M|^{m-1} \left|M^{1,\ldots,m}_{1,\ldots,m}\right|. \quad (3.33)$$

Prove this equation. Show that it implies both equation (3.29) and (3.30).

3.5.4 Use the Desnanot–Jacobi adjoint matrix theorem to prove the Vandermonde formula by induction on the number of variables.

3.5.5 Prove that if we define the λ-determinant by equation (3.31), then the ordinary determinant is what we get when $\lambda = 1$.

3.5.6 Use the inductive definition of the λ-determinant to prove that

$$\left|x_j^{n-i}\right|_\lambda = \prod_{1\le i<j\le n} (x_i + \lambda x_j).$$

3.5.7 Given an alternating sign matrix B, let $B(j)$ be the number of times that j appears in the monotone triangle that corresponds to B. Prove that equation (3.32) can also be written as

$$\prod_{1\le i<j\le n} (x_i + \lambda x_j) = \sum_{B\in\mathcal{A}_n} \lambda^{\mathcal{I}(B)}(1+\lambda^{-1})^{N(B)} \prod_{j=1}^{n} x_j^{B(j)-1}. \tag{3.34}$$

3.5.8 Verify equation (3.32) for $n = 4$.

3.5.9 Using equation (3.32), prove that

$$\sum_{B\in\mathcal{A}_n} x^{\mathcal{I}(B)}(1+x^{-1})^{N(B)} = (1+x)^{n(n-1)/2}.$$

3.5.10 Using equation (3.32), prove that

$$\sum_{B\in\mathcal{A}_n} 2^{N(B)} = 2^{n(n-1)/2}.$$

4
Symmetric Functions

The reader must be warned that, although there is little doubt that this result [the generating function for symmetric plane partitions] is correct, its truth at present rests upon the fact that it represents faithfully every particular case that has been examined, and that some consequences deduced from it also appear to be correct. The result has not been rigorously established.... Further investigations in regard to these matters would be sure to lead to valuable work.

– Percy A. MacMahon (1916, p. 267)

Plane partitions might have remained nothing more than a mathematical curiosity were it not for the rise of a field within modern algebra known as **representation theory**. Given a group such as $\mathcal{S}_n$, the symmetric group on n letters, a **representation** of this group is a map from the group to a set of matrices, a map that preserves the group action. For $\mathcal{S}_n$, we have already seen such a map; it is the one that takes each permutation to the corresponding permutation matrix.

Recall that multiplication of permutation matrices exactly follows composition of permutations. For example, if σ is defined by the sequence 45312,

$$\sigma(1) = 4,\ \sigma(2) = 5,\ \sigma(3) = 3,\ \sigma(4) = 1,\ \sigma(5) = 2,$$

and τ is defined by 35124, then the composition $\sigma \circ \tau$ is described by 32451,

$$\sigma(\tau(1)) = \sigma(3) = 3, \quad \sigma(\tau(2)) = \sigma(5) = 2, \ \ldots\ .$$

This relationship is maintained (though multiplication is taken in reverse order) when we look at the corresponding matrices:

$$\begin{pmatrix} 0 & 0 & 1 & 0 & 0 \\ 0 & 0 & 0 & 0 & 1 \\ 1 & 0 & 0 & 0 & 0 \\ 0 & 1 & 0 & 0 & 0 \\ 0 & 0 & 0 & 1 & 0 \end{pmatrix} \begin{pmatrix} 0 & 0 & 0 & 1 & 0 \\ 0 & 0 & 0 & 0 & 1 \\ 0 & 0 & 1 & 0 & 0 \\ 1 & 0 & 0 & 0 & 0 \\ 0 & 1 & 0 & 0 & 0 \end{pmatrix} = \begin{pmatrix} 0 & 0 & 1 & 0 & 0 \\ 0 & 1 & 0 & 0 & 0 \\ 0 & 0 & 0 & 1 & 0 \\ 0 & 0 & 0 & 0 & 1 \\ 1 & 0 & 0 & 0 & 0 \end{pmatrix}.$$

Up to isomorphism, this is the only one-to-one representation of $\mathcal{S}_5$, but $\mathcal{S}_5$ does have other representations. We could map every element to the 1×1 matrix 1. Slightly more interesting is the representation that takes a permutation to $+1$ if its inversion number is even and to -1 if its inversion number is odd. To say that this is a representation is equivalent to the statement that

$$(-1)^{\mathcal{I}(\sigma \circ \tau)} = (-1)^{\mathcal{I}(\sigma)}(-1)^{\mathcal{I}(\tau)}.$$

The study of the symmetric group has always been closely related to the study of the symmetric functions that were defined in the introduction to Chapter 2. In fact, much of group theory arose as an abstraction of the study of symmetric functions. In the decades spanning 1900, symmetric functions reappeared as powerful tools in the analysis of representations of $\mathcal{S}_n$. We have seen that symmetric functions are related to the counting of partitions. In the early 1900s, Alfred Young pioneered the use of planar objects, what are today known as Young tableaux, to study representations of the symmetric group and derive identities in the group algebra. At the same time, Issai Schur and Georg Frobenius were demonstrating how to use a family of symmetric functions that had been known to Cauchy but which today we call Schur functions to study **characters** of representations, sums of the diagonal elements in the matrix representation.

Young tableaux can be viewed as plane partitions; Schur functions as sums of monomials indexed by column strict plane partitions. The machinery of the theory of representations of the symmetric group provides insight into plane partitions. It is from this vantage that Ian Macdonald discovered his proof of MacMahon's conjecture (Conjecture 4) for the generating function for symmetric plane partitions inside $\mathcal{B}(r, r, t)$.

4.1 Schur functions

The original impetus for studying symmetric functions came from the problem of finding a formula for the roots of a polynomial, a formula that could be expressed in terms of the coefficients of that polynomial using the four arithmetic operations and the taking of square roots. The quadratic formula does this for polynomials of degree two. Formulæ for cubic and biquadratic (degree four) polynomials were discovered in the 1500s by Scipione del Ferro, Niccolò Tartaglia, Lodovico Ferrari, and Gerolamo Cardano. By the late 1700s, it was beginning to appear that no such formula existed for the general quintic polynomial, a result that was finally proven by Niels Henrik Abel in 1824. But there are specific polynomials for which there are such formulæ. The procedure for deciding whether or not such a formula exists and deriving it when it does is what Galois theory is all about.

The key to tackling this problem was the realization – going back at least to François Viète in the 1500s – that the coefficients of a monic polynomial with distinct roots are, up to sign, the elementary symmetric functions in those roots. The **elementary symmetric function**, $e_k(x_1, x_2, \ldots, x_n)$, is the monomial symmetric function (see page 34) that corresponds to the partition with k 1s:

$$\begin{aligned} e_0 &= 1, \\ e_1 &= x_1 + x_2 + \cdots + x_n, \\ e_2 &= x_1x_2 + x_1x_3 + \cdots + x_{n-1}x_n, \\ &\vdots \\ e_k &= \sum_{1 \le i_1 < i_2 < \cdots < i_k \le n} x_{i_1} x_{i_2} \cdots x_{i_k}. \end{aligned}$$

Viète realized that

$$\prod_{i=1}^{n} (x - x_i) = x^n - e_1\, x^{n-1} + e_2\, x^{n-2} - \cdots + (-1)^n e_n.$$

This is equivalent to the generating function identity that can be used to define the elementary symmetric functions:

$$\prod_{j=1}^{n} (1 + x_j t) = \sum_{i=0}^{\infty} e_i(x_1, x_2, \ldots, x_n)\, t^i. \tag{4.1}$$

Note that for i larger than n, $e_i(x_1, \ldots, x_n)$ is zero.

Elementary symmetric functions as basis

As we saw in the introduction to Chapter 2, each homogeneous symmetric function of degree n is a linear combination of the monomial symmetric functions indexed by the partitions of n. Another way of saying this is that the monomial symmetric functions form a **basis** of order $p(n)$ for the space of all homogeneous symmetric functions of degree n. In 1762, Edward Waring proved that we can also construct a basis out of products of the elementary symmetric functions. This implies that any symmetric function in the roots of a polynomial can be expressed as a sum of products of the coefficients of that polynomial.

Given a partition, λ, we define the corresponding basis element by

$$e_\lambda = e_{(\lambda_1,\lambda_2,\ldots,\lambda_n)} = e_{\lambda_1} e_{\lambda_2} \cdots e_{\lambda_n}.$$

The five basis elements for the homogeneous symmetric functions of degree four in four variables, x_1, x_2, x_3, x_4, are

$$\begin{aligned}
e_4 &= x_1x_2x_3x_4, \\
e_{(3,1)} &= (x_1x_2x_3 + x_1x_2x_4 + x_1x_3x_4 + x_2x_3x_4) \\
&\quad \times (x_1 + x_2 + x_3 + x_4) \\
e_{(2,2)} &= (x_1x_2 + x_1x_3 + x_1x_4 + x_2x_3 + x_2x_4 + x_3x_4)^2, \\
e_{(2,1,1)} &= (x_1x_2 + x_1x_3 + x_1x_4 + x_2x_3 + x_2x_4 + x_3x_4) \\
&\quad \times (x_1 + x_2 + x_3 + x_4)^2, \\
e_{(1,1,1,1)} &= (x_1 + x_2 + x_3 + x_4)^4.
\end{aligned}$$

Proposition 4.1 *The set of e_λ where λ ranges over all partitions of n is a basis for the homogeneous symmetric functions of degree n.*

Proof: We first define a complete order on all partitions of n. Given partitions $\lambda = (\lambda_1 \geq \lambda_2 \geq \cdots \geq \lambda_n \geq 0)$ and $\mu = (\mu_1 \geq \mu_2 \geq \cdots \geq \mu_n \geq 0)$, we find the smallest i such that $\lambda_i \neq \mu_i$. We consider λ to be larger than μ if and only if $\lambda_i > \mu_i$. For example, the partitions of 7 are ordered by

$$\begin{gathered}
(7) > (6,1) > (5,2) > (5,1,1) > (4,3) > \\
> (4,2,1) > (4,1,1,1) > (3,3,1) > (3,2,2) > (3,2,1,1) > \cdots.
\end{gathered}$$

Given a partition λ, we let λ' denote its conjugate (see page 44). We observe that e_λ equals $m_{\lambda'}$ plus a sum of monomial symmetric functions indexed by partitions that are strictly less than λ'. As an example, the basis elements of degree four are

$$\begin{aligned}
e_4 &= m_{(1,1,1,1)}, \\
e_{(3,1)} &= m_{(2,1,1)} + 4m_{(1,1,1,1)},
\end{aligned}$$

$$
\begin{aligned}
e_{(2,2)} &= m_{(2,2)} + 2m_{(2,1,1)} + 6m_{(1,1,1,1)},\\
e_{(2,1,1)} &= m_{(3,1)} + 2m_{(2,2)} + 5m_{(2,1,1)} + 12m_{(1,1,1,1)},\\
e_{(1,1,1,1)} &= m_{(4)} + 4m_{(3,1)} + 6m_{(2,2)} + 12m_{(2,1,1)} + 24m_{(1,1,1,1)}.
\end{aligned}
$$

Since the monomial symmetric functions give us a basis for the homogeneous symmetric functions of degree n, we only need to show that each monomial symmetric function can be expressed as a sum of elementary symmetric functions. This is clearly true for $m_{(1^n)} = e_n$. We now proceed by induction, moving up the order on the partitions of n. We assume that for the partition λ, if $\mu < \lambda$, then m_μ can be written as a sum of elementary symmetric functions. From our observation, we know that m_λ equals $e_{\lambda'}$ minus a sum of monomial symmetric functions indexed by partitions less than λ. Since each of these other monomial symmetric functions is a linear combination of elementary symmetric functions, so must be m_λ.

Q.E.D.

Complete symmetric functions

A third basis is given by the **complete symmetric functions**. We define h_i to be the sum of all monomial symmetric functions of degree i, and then extend this in the same way we did for elementary symmetric functions:

$$h_{(\lambda_1,\lambda_2,\ldots,\lambda_n)} = h_{\lambda_1} h_{\lambda_2} \cdots h_{\lambda_n}.$$

The five complete symmetric functions of degree four (in four variables) are

$$
\begin{aligned}
h_4 &= m_{(4)} + m_{(3,1)} + m_{(2,2)} + m_{(2,1,1)} + m_{(1,1,1,1)},\\
h_{(3,1)} &= (m_{(3)} + m_{(2,1)} + m_{(1,1,1)})m_{(1)},\\
h_{(2,2)} &= (m_{(2)} + m_{(1,1)})^2,\\
h_{(2,1,1)} &= (m_{(2)} + m_{(1,1)})m_{(1)}^2,\\
h_{(1,1,1,1)} &= m_{(1)}^4.
\end{aligned}
$$

The generating function for the complete symmetric functions is

$$\sum_{i=0}^{\infty} h_i(x_1, x_2, \ldots, x_n)\, t^i = \prod_{j=1}^{n} \frac{1}{1 - x_j t}. \tag{4.2}$$

Schur functions

While both the elementary and complete symmetric functions will have a role to play in this chapter, the basis that most interests us is one that is defined in terms of determinants, the **Schur functions**, s_λ, where

$$\begin{aligned} s_{(\lambda_1,\lambda_2,\ldots,\lambda_n)}(x_1,\ldots,x_n) &= \frac{\det(x_j^{n-i+\lambda_i})}{\det(x_j^{n-i})} \\ &= \frac{\det(x_j^{n-i+\lambda_i})}{\prod_{1\le i<j\le n}(x_i-x_j)}. \end{aligned} \tag{4.3}$$

As an example, we take $n = 3$ and look at the Schur functions for the partitions

$\lambda = (2,1,1)$:

$$\begin{aligned} s_{2,1,1}(x,y,z) &= \frac{\begin{vmatrix} x^4 & y^4 & z^4 \\ x^2 & y^2 & z^2 \\ x & y & z \end{vmatrix}}{(x-y)(x-z)(y-z)} \\ &= \frac{x^4y^2z - x^2y^4z - x^4yz^2 + xy^4z^2 + x^2yz^4 - xy^2z^4}{x^2y - xy^2 - x^2z + y^2z + xz^2 - yz^2} \\ &= x^2yz + xy^2z + xyz^2. \end{aligned}$$

$\lambda = (4,1,0)$:

$$\begin{aligned} s_{4,1,0}(x,y,z) &= \frac{\begin{vmatrix} x^6 & y^6 & z^6 \\ x^2 & y^2 & z^2 \\ 1 & 1 & 1 \end{vmatrix}}{(x-y)(x-z)(y-z)} \\ &= x^4y + x^4z + xy^4 + y^4z + xz^4 + yz^4 \\ &\quad + x^3y^2 + x^3z^2 + x^2y^3 + y^3z^2 + x^2z^3 + y^2z^3 \\ &\quad + 2x^3yz + 2xy^3z + 2xyz^3 \\ &\quad + 2x^2y^2z + 2x^2yz^2 + 2xy^2z^2. \end{aligned}$$

If λ is a partition of m, then the determinant in the numerator of $s_\lambda(x_1,\ldots,x_n)$ is an alternating polynomial of degree $m + n(n-1)/2$. From Proposition 2.8 on page 65, we know that $s_\lambda(x_1,\ldots,x_n)$ is a symmetric polynomial of degree m.

Schur functions are named for Issai Schur (1875–1941), who used them in his work in invariant theory, but they are much older. In his 1815 paper, Cauchy described these functions and proved that they are symmetric polynomials. In 1841, Jacobi (1841c) showed how to express these polynomials in terms of the complete symmetric polynomials, the Jacobi–Trudi identity which we now state. Nicolas Trudi (1811–1884) was a Neapolitan mathematician who gave a simplified proof of this identity in 1864.

Proposition 4.2 (The Jacobi–Trudi identity) *Let $\lambda = (\lambda_1, \ldots, \lambda_n)$ be a partition into at most n parts. We have that*

$$s_\lambda(x_1, \ldots, x_n) = \det\left(h_{\lambda_i+j-i}\right)_{i,j=1}^{k}. \tag{4.4}$$

Proof: Our proof uses the generating functions for the elementary and complete symmetric functions, equations (4.1) and (4.2). Let $\alpha = (\alpha_1, \alpha_2, \ldots, \alpha_k)$ be any sequence of non-negative integers, and define the $k \times k$ matrices

$$A_\alpha = (x_j^{\alpha_i}), \qquad H_\alpha = (h_{\alpha_i-k+j}), \qquad M = \left((-1)^{k-i} e_{k-i}^{(j)}\right),$$

where $e_{k-i}^{(j)}$ denotes the $(k-i)$th elementary symmetric function in all of the variables *except* x_j. We shall prove that for any α, $A_\alpha = H_\alpha M$, and therefore

$$\det(H_\alpha) = \det(A_\alpha) \,/\, \det(M). \tag{4.5}$$

We begin with the generating functions identities for the complete and elementary symmetric functions:

$$\begin{aligned}\sum_{m=0}^{\infty} h_m t^m \sum_{n=0}^{k-1} e_n^{(l)} (-t)^n &= \prod_{m=1}^{k} \frac{1}{1-x_m t} \prod_{\substack{n=1\\ n\neq l}}^{k} (1-x_n t)\\ &= \frac{1}{1-x_l t} = 1 + x_l t + x_l^2 t^2 + \cdots.\end{aligned}$$

When we compare the coefficient of t^{α_i} on each side of this equation, we see that

$$\sum_{j=1}^{k} h_{\alpha_i-k+j} (-1)^{k-j} e_{k-j}^{(l)} = x_l^{\alpha_i}.$$

This is the same as saying that $H_\alpha M = A_\alpha$.

When $\alpha = (k-1, k-2, \ldots, 0)$, H_α has 1s on the main diagonal and 0s below the main diagonal, so its determinant is 1. This implies that

$$\det(M) = \det\left(x_j^{k-i}\right).$$

Equation (4.4) is the special case of equation (4.5) in which

$$\alpha = (\lambda_1 + k - 1, \lambda_2 + k - 2, \ldots, \lambda_k).$$

Q.E.D.

Exercises

4.1.1 What are the partitions of 29 that immediately precede and follow $7 + 6 + 6 + 4 + 4 + 2$ in descending order?

4.1.2 Calculate $e_3(1, 1, 1, 1, 1)$. What is the value of e_n when there are m variables that are each set equal to 1? What is the value of $h_3(1, 1, 1, 1, 1)$? What is the value of h_n when there are m variables that are each set equal to 1?

4.1.3 Prove that if σ and τ are any two permutations on n letters, then

$$(-1)^{\mathcal{I}(\sigma\circ\tau)} = (-1)^{\mathcal{I}(\sigma)}(-1)^{\mathcal{I}(\tau)}.$$

4.1.4 Prove that $(-1)^{\mathcal{I}(\sigma^{-1})} = (-1)^{\mathcal{I}(\sigma)}$.

4.1.5 Find the expansions of the five monomial symmetric functions of degree 4 in terms of the elementary symmetric functions.

4.1.6 Prove that

$$h_r(1, q, q^2, \ldots, q^n) = \begin{bmatrix} n+r \\ r \end{bmatrix}.$$

4.1.7 Prove that

$$e_r(1, q, q^2, \ldots, q^{n-1}) = q^{r(r-1)/2} \begin{bmatrix} n \\ r \end{bmatrix}.$$

4.1.8 Prove that

$$e_n = \det(h_{1-i+j})_{i,j=1}^n \quad \text{and} \tag{4.6}$$

$$h_n = \det(e_{1-i+j})_{i,j=1}^n. \tag{4.7}$$

The first of these equalities implies that the complete homogeneous symmetric functions of degree n form a basis for the elementary symmetric functions of degree n. It follows that they can be used as a basis for all homogeneous symmetric functions of degree n.

4.1.9 Use equations (4.4) and (4.6) to prove that if we work with symmetric functions in k variables, then

$$s_{a+1,1^b} = h_{a+1}e_b - h_{a+2}e_{b-1} + h_{a+3}e_{b-2} - \cdots + (-1)^{n-1}h_{a+k},$$

where $\lambda = a+1, 1^b$ is the partition with one part of size $a+1$ and b parts of size 1.

4.2 Semistandard tableaux

The coefficients of the monomials in any Schur function are always non-negative integers. They count something. We shall prove that the coefficient of $x_1^{a_1}x_2^{a_2}\cdots x_n^{a_n}$ in $s_\lambda(x_1,\ldots,x_n)$ is the number of ways in which we can replace the dots in the Ferrers diagram of λ with our variables $x_1,\ldots,x_n$ so that for each i we have a_i copies of x_i and such that the subscripts are weakly increasing as we move from left to right across any row and the subscripts are strictly increasing as we move down any column. For example, the coefficient of $x_1^2x_2^2x_3^2x_4^4x_5^2$ in $s_{(4,4,3,1)}$ is 7. One of the seven arrangements is

$$\begin{array}{cccc} x_1 & x_1 & x_2 & x_3 \\ x_2 & x_3 & x_4 & x_4 \\ x_4 & x_4 & x_5 & \\ x_5 & & & \end{array}.$$

I invite the reader to find the other six.

For simplicity, we usually record just the subscripts in what is called a **semistandard tableau of shape** λ. This is the Ferrers diagram of λ in which the dots have been replaced by positive integers with the restriction that these integers are weakly increasing across rows and strictly increasing down columns. An example of a semistandard tableau of shape $(5,3,3,2,1,1)$ with $n=6$ is

$$\begin{array}{ccccc} 1 & 1 & 3 & 3 & 4 \\ 2 & 3 & 4 & & \\ 3 & 5 & 5 & & \\ 4 & 6 & & & \\ 5 & & & & \\ 6 & & & & \end{array}$$

Given a tableau T, we will denote the corresponding monomial by $\mathbf{x}^T$. In this case $\mathbf{x}^T = x_1^2x_2x_3^4x_4^3x_5^3x_6^2$. We are now ready for our second definition of the Schur function.

Theorem 4.3 *An equivalent definition of the Schur function is given by*

$$s_\lambda(x_1, x_2, \ldots, x_n) = \sum_T \mathbf{x}^T, \tag{4.8}$$

where the sum is over all semistandard tableaux of shape λ with entries chosen from $\{1, 2, \ldots, n\}$.

This representation makes it easier to describe the Schur function associated with a particular partition, though it is not at all obvious from this definition that the Schur function must be symmetric.

Proof of Theorem 4.3

We shall prove that

$$\sum_{T \in \mathcal{T}_{\lambda,n}} \mathbf{x}^T = \det \left(h_{\lambda_i + j - i}\right)_{i,j=1}^{k}, \tag{4.9}$$

where $\mathcal{T}_{\lambda,n}$ is the set of all semistandard tableaux of shape λ with entries taken from the set $\{1, 2, \ldots, n\}$. The theorem then follows from the Jacobi–Trudi identity.

We specify a partition $\lambda = (\lambda_1 \geq \lambda_2 \geq \cdots \geq \lambda_k \geq 0)$ and a positive integer n. There is a natural way of representing each tableau in $\mathcal{T}_{\lambda,n}$ as a nest of lattice paths. Let T_{ij} be the entry in row i, column j of the semistandard tableaux T. The ith lattice path (counted from the top) encodes the ith row of the tableau. It goes from $(1, k-i)$ to $(n, k+\lambda_i - i)$. It makes λ_i steps to the north, and the jth step to the north is made along the line $x = T_{ij}$.

As an example, the semistandard tableau

$$\begin{array}{ccccc} 1 & 1 & 2 & 4 & 5 \\ 2 & 3 & 3 & 5 & \\ 4 & 4 & 4 & & \\ 5 & 6 & 6 & & \\ 6 & & & & \end{array}$$

corresponds to the nest of lattice paths given in Figure 4.1. It has weight $x_1^2\, x_2^2\, x_3^2\, x_4^4\, x_5^3\, x_6^3$.

We have established a one-to-one correspondence between the semistandard tableaux in $\mathcal{T}_{\lambda,n}$ and sets of non-intersecting lattice paths with the specified start and end points. The condition that each column

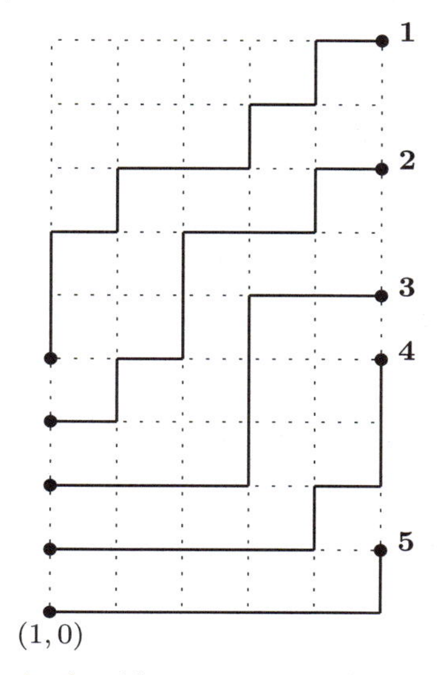

Figure 4.1. A semistandard tableau represented as a nest of lattice paths. Vertical steps along the line $x = j$ represent occurrences of the letter j in the tableau.

of the tableau is strictly increasing is equivalent to the statement that the paths are non-intersecting. The power of x_j in the weight of any particular nest of paths is the number of steps to the north taken along the vertical line $x = j$.

We now play the same sort of trick that we used in Chapter 3. We take all nests of lattice paths in which each path starts at one of the points $(1, k - j)$ and ends at one of the points $(n, k + \lambda_i - i)$, insisting that no two paths share the same start or end points, but allowing the paths to intersect. As before, this defines a permutation, σ. The ith path starts at $(1, k - \sigma(i))$ and ends at $(n, k + \lambda_i - i)$. Its weight is the product over j of x_j to a power equal to the number of vertical steps along the line $x = j$. The weight of the ith path is a monomial of degree $\lambda_i + \sigma(i) - i$. The sum over all possible lattice paths from $(1, k - \sigma(i))$ to $(n, k + \lambda_i - i)$ is just the sum of all possible monomials of degree $\lambda_i + \sigma(i) - i$ in the variables $x_1, x_2, \ldots, x_n$. This is the complete symmetric function $h_{\lambda_i + \sigma(i) - i}$.

The weight of a nest of lattice paths is the product of the weights of the individual paths, multiplied by $(-1)^{\mathcal{I}(\sigma)}$. It follows that the sum over all possible nests of lattice paths with these particular starting and ending points is

$$\sum_{\sigma \in \mathcal{S}_k} (-1)^{\mathcal{I}(\sigma)} \prod_{i=1}^{k} h_{\lambda_i+\sigma(i)-i} = \det\left(h_{\lambda_i+j-i}\right)_{i,j=1}^{k},$$

where we define $h_0 = 1$ and $h_m = 0$ if m is a negative integer.

Exactly as in Chapter 3, we can pair up and cancel all of the intersecting paths. As before, we switch two tails. This changes the parity of the inversion number of the permutation, but it does not change the location of the vertical steps, so it does not change the corresponding monomial. This means that the sum over all possible nests of lattice paths is equal to the sum over all nests of non-intersecting paths (for which the permutation *must* be the identity). This proves equation (4.9).

Q.E.D.

Using Schur functions

Schur functions give us a lot of flexibility in constructing generating functions for plane partitions. If we rewrite the tableau on page 127 in terms of the variables, it is

$$\begin{array}{ccccc} x_1 & x_1 & x_3 & x_3 & x_4 \\ x_2 & x_3 & x_4 & & \\ x_3 & x_5 & x_5 & & \\ x_4 & x_6 & & & \\ x_5 & & & & \\ x_6 & & & & \end{array}$$

Each of these six variables can represent stacks of cubes of different heights. For example, we might want to set $x_1 = q^6$: Each x_1 represents a stack of six cubes. If we set $x_2 = q^5$, $x_3 = q^4$, $x_4 = q^3$, $x_5 = q^2$, and $x_6 = q$, then our example represents the plane partition given in Figure 4.2. The power of q in the corresponding monomial will be the total number of cubes in the plane partition.

This implies that $s_{(5,3,3,2,1,1)}(q^6, q^5, q^4, q^3, q^2, q)$ is the generating function for column strict plane partitions with all stack heights less than or equal to 6 and with row lengths given by the parts of λ: 5, 3, 3, 2, 1, and 1. We shall use this fact to give a simple proof of MacMahon's generating function for plane partitions that fit inside an $r \times s \times t$ box

Figure 4.2. The plane partition defined by the sample tableau.

(Theorem 1.3):

$$\prod_{i=1}^{r}\prod_{k=1}^{t}\frac{1-q^{i+k+s-1}}{1-q^{i+k-1}}.$$

In the next section, this idea will provide the foundation on which we prove MacMahon's conjecture.

Proof of the generating function for plane partitions

Let $\lambda = s^r$ be the partition with r copies of s. The Schur function s_λ evaluated at $x_1 = q^{t+r}, x_2 = q^{t+r-1}, \ldots, x_{t+r} = q$ is the generating function for column strict (strict decrease down columns) plane partitions with exactly r rows, each of length s, with the largest stack of height less than or equal to $t + r$. If we remove one cube from each stack in row r, two cubes from each stack in row $r - 1$, and so on up to r cubes from each stack in the first row, then we are left with precisely what we want: a plane partition that is a subset of $\mathcal{B}(r, s, t)$. Since each plane partition inside $\mathcal{B}(r, s, t)$ can be obtained uniquely in this manner, the generating function that we seek is

$$q^{-sr(r+1)/2} s_\lambda(q^{t+r}, \ldots, q), \tag{4.10}$$

where

$$\lambda_i = \begin{cases} s, & \text{if } 1 \leq i \leq r, \\ 0, & \text{if } r < i \leq t + r. \end{cases}$$

We use the definition of the Schur function to write $s_\lambda(q^{t+r}, \ldots, q)$ as a quotient of determinants: X/Y where

$$\begin{aligned} X &= \det\left((q^{t+r-j+1})^{t+r-i+\lambda_i}\right)_{i,j=1}^{t+r}, \\ Y &= \prod_{1\le i<j\le t+r} (q^{t+r-i+1} - q^{t+r-j+1}). \end{aligned}$$

We begin by simplifying X. For each i, we take a common factor of $q^{t+r-i+\lambda_i}$ out of the ith row. This gives us

$$X = q^{sr+(t+r)(t+r-1)/2} \det\left(q^{(t+r-j)(t+r-i+\lambda_i)}\right).$$

We can simplify this using the Vandermonde formula,

$$X = q^{sr+(t+r)(t+r-1)/2} \prod_{1\le i<j\le t+r} \left(q^{t+r-i+\lambda_i} - q^{t+r-j+\lambda_j}\right). \qquad (4.11)$$

We take a factor of q out of each of the $(t+r)(t+r-1)/2$ terms in Y. This yields

$$Y = q^{(t+r)(t+r-1)/2} \prod_{1\le i<j\le t+r} (q^{t+r-i} - q^{t+r-j}). \qquad (4.12)$$

We put our representation for the generating function, equation (4.10), together with equations (4.11) and (4.12). This establishes that our generating function is equal to

$$q^{-sr(r-1)/2} \prod_{1\le i<j\le t+r} \frac{q^{t+r-i+\lambda_i} - q^{t+r-j+\lambda_j}}{q^{t+r-i} - q^{t+r-j}}.$$

When i and j are both less than or equal to r, the quantity inside the product is

$$\frac{q^{t+r-i+s} - q^{t+r-j+s}}{q^{t+r-i} - q^{t+r-j}} = q^s.$$

There are $r(r-1)/2$ such pairs (i,j), and so this cancels the power of q in front of the product. When i and j are both larger than r, the quantity inside the product is 1. The only terms that are now left are those for which $1 \le i \le r$ and $r+1 \le j \le r+t$, and therefore the generating function is

$$\prod_{i=1}^{r} \prod_{j=r+1}^{r+t} \frac{q^{t+r-i+s} - q^{t+r-j}}{q^{t+r-i} - q^{t+r-j}}.$$

We multiply numerator and denominator of each term by q^{j-t-r}:

$$\prod_{i=1}^{r} \prod_{j=r+1}^{r+t} \frac{1-q^{j-i+s}}{1-q^{j-i}}.$$

We now replace the index j by $k+r$ and the index i by $r+1-i$ to put this in the desired form.

Q.E.D.

Exercises

4.2.1 Find the other six arrangements of $x_1^2\, x_2^2\, x_3^2\, x_4^4\, x_5^2$ in the Ferrers diagram for $\lambda = (4,4,3,1)$.

4.2.2 Use Theorem 4.3 to find the coefficient of $x_1^2\, x_2^3\, x_3\, x_4^4\, x_5^3$ in the Schur function $s_{(5,4,3,1)}(x_1,x_2,x_3,x_4,x_5)$.

4.2.3 Use Theorem 4.3 to prove that s_λ equals m_λ plus a sum of monomial symmetric functions indexed by partitions smaller than λ. Use this fact to prove that the Schur functions of degree n form a basis for all homogeneous symmetric functions of degree n.

4.2.4 Prove that if λ is the partition $\lambda_1 \geq \lambda_2 \geq \cdots \geq \lambda_r$, then $s_\lambda(q^n, q^{n-1}, \ldots, q)$ is equal to the generating function for all plane partitions whose bottom level is the Ferrers graph for λ, for which the stack in the ith row and jth column is strictly shorter than the stack in the $i-1$st row and jth column, and that has no stack with more than n cubes.

4.2.5 Let λ_i' be the ith largest part in λ', the partition conjugate to λ. Let k be the number of parts in λ'. Prove that

$$s_\lambda = \det(e_{\lambda_i'-i+j})_{i,j=1}^{k}.$$

4.2.6 In the next section, we shall prove that

$$\sum_{\lambda} s_\lambda(x_1,\ldots,x_n) = \prod_{i=1}^{n} \frac{1}{1-x_i} \prod_{1\leq i<j\leq n} \frac{1}{1-x_i x_j}, \tag{4.13}$$

where the summation is over all partitions, λ, with at most n parts. Use equation (4.13) to prove that the generating function for column strict plane partitions is given by

$$\prod_{i=1}^{\infty} \frac{1}{1-q^i} \prod_{1\leq i<j} \frac{1}{1-q^{i+j}} = \prod_{n=1}^{\infty} \frac{1}{(1-q^n)^{\lceil n/2\rceil}},$$

where $\lceil n/2 \rceil$, read the **ceiling** of $n/2$, is the smallest integer greater than or equal to $n/2$.

4.2.7 Use equation (4.13) to prove that the generating function for column strict plane partitions in which each stack has odd height is given by

$$\prod_{i=1}^{\infty} \frac{1}{1-q^{2i-1}} \prod_{1 \le i < j} \frac{1}{1-q^{2i+2j-2}} = \prod_{n=1}^{\infty} \frac{1}{(1-q^n)^{\nu(n)}},$$

where $\nu(n) = 1$ if n is odd and $\nu(n) = \lfloor n/4 \rfloor$ if n is even.

4.2.8 Use equation (4.13) to prove the general result of Bender and Knuth (1972) that if S is any set of positive integers, then the generating function for column strict plane partitions in which each stack height is an element of S is given by

$$\prod_{i \in S} \frac{1}{1-q^i} \prod_{\substack{i,j \in S \\ i<j}} \frac{1}{1-q^{i+j}}.$$

4.3 Proof of the MacMahon conjecture

The proof of MacMahon's conjecture (Conjecture 4) is the first of three major proofs we shall study. It was a watershed in the history of plane partitions and of algebraic combinatorics, a dramatic example of the application of tools developed for algebraic studies to a problem that had been posed as purely combinatorial.

Macdonald's proof appeared in *Symmetric Functions and Hall Polynomials* (1979), a book that consolidated much of what was known, advanced new understandings, and suggested many research problems. It is difficult to assess cause and effect, but its publication coincided with increased activity at the intersection of algebra and combinatorics. Neither this chapter nor this book does justice to algebraic combinatorics. Its methods are more sophisticated than anything presented here. Nevertheless, the proofs of MacMahon's conjecture and of the refined alternating sign matrix conjecture hint at the power and excitement of this field of exploration.

We recall that a plane partition is symmetric when there is a cube at position (i, j, k) if and only if there is a cube at position (j, i, k) (see Fig. 1.5 on page 13). Ian Macdonald and George Andrews first entered our story with their proofs of MacMahon's conjecture that the generating function for symmetric plane partitions that are contained in $\mathcal{B}(r, r, t)$

is $\prod_{\eta \in \mathcal{B}(r,r,t)/\mathcal{S}_2}(1 - q^{|\eta|(1+\mathrm{ht}(\eta))})/(1 - q^{|\eta|\,\mathrm{ht}(\eta)})$, which is the same as (exercise 1.2.8, page 17)

$$\prod_{i=1}^{r} \frac{1 - q^{t+2i-1}}{1 - q^{2i-1}} \prod_{1 \leq i < j \leq r} \frac{1 - q^{2(t+i+j-1)}}{1 - q^{2(i+j-1)}}. \tag{4.14}$$

In the late 1960s, Basil Gordon (1971b) proved the special case of MacMahon's conjecture in which we take $|q| < 1$ and let the vertical bound, t, approach infinity. The first step in his proof is also the first step in Macdonald's. It is to express the generating function for symmetric plane partitions in terms of Schur functions.

Lemma 4.4 *The generating function for symmetric plane partitions that are subsets of $\mathcal{B}(r, r, t)$ is*

$$\sum_{\lambda \subseteq \{t^r\}} s_\lambda(q^{2r-1}, q^{2r-3}, \ldots, q^3, q), \tag{4.15}$$

where the summation is over all partitions λ with at most r parts, each of which is less than or equal to t.

For the second step, we shall prove Macdonald's formula for the sum of all Schur functions whose partitions, λ, fit inside an $r \times t$ rectangle.

Lemma 4.5 *The sum of Schur functions, s_λ, taken over all partitions into at most r parts, each of which is less than or equal to t, satisfies the equation*

$$\sum_{\lambda \subseteq \{t^r\}} s_\lambda(x_1, \ldots, x_r) = \frac{\det(x_i^{j-1} - x_i^{t+2r-j})}{\det(x_i^{j-1} - x_i^{2r-j})}. \tag{4.16}$$

The final part of the proof is to use the Weyl denominator formula for $\mathbf{B_n}$, Theorem 2.10 on page 68, to rewrite both determinants as products. The hardest part of the proof of MacMahon's conjecture is Lemma 4.5. Macdonald derived it from a general identity for a multi-parameter family of symmetric functions known as the Hall–Littlewood polynomials. Since we need only this special case, we shall take a much more direct approach and prove it by induction on the number of variables.

Proof of Lemma 4.4

We need to prove that for each integer n there is a one-to-one correspondence between symmetric plane partitions of n that are contained

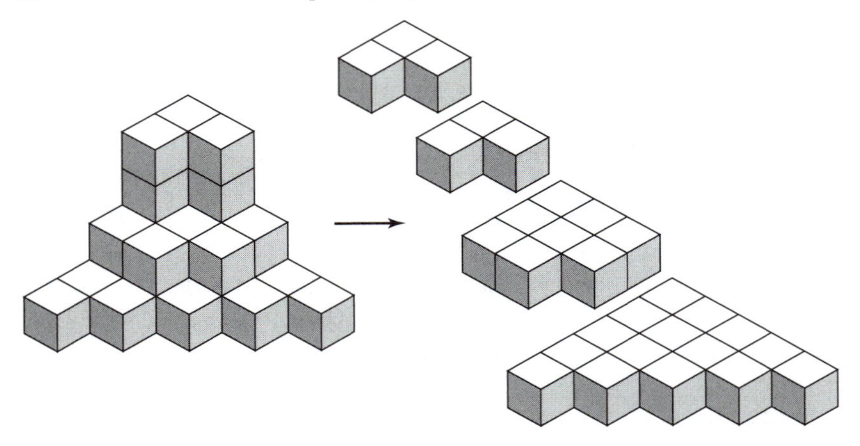

Figure 4.3. Decomposing a symmetric plane partition.

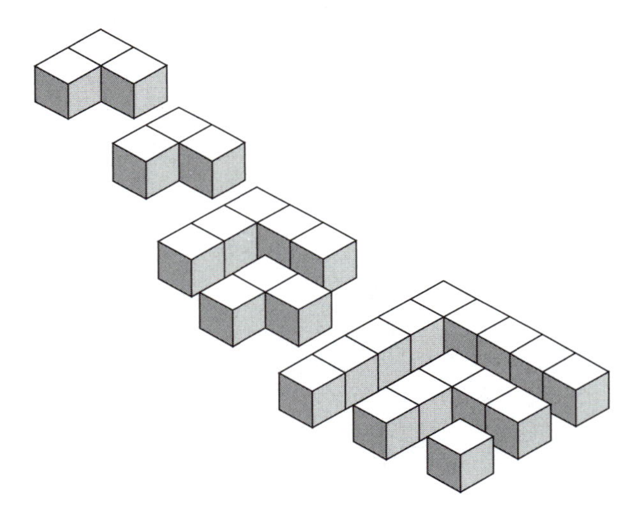

Figure 4.4. Decomposing a symmetric plane partition into angles.

in $\mathcal{B}(r, r, t)$ and column strict plane partitions of n with stacks of odd height contained in $\mathcal{B}(r, t, 2r-1)$. The latter set of plane partitions is counted by $s_\lambda(q^{2r-1}, q^{2r-3}, \ldots, q^3, q)$ (see exercise 4.2.7 on page 134).

We begin with a symmetric plane partition and decompose it into **levels**, subsets of points with the same height (see Fig. 4.3). Each level is itself symmetric about the $x = y$ plane, and naturally decomposes into **angles** (see Fig. 4.4). The first angle consists of cubes that lie in either

Figure 4.5. Plane partition with strict decrease down columns and stacks of odd height.

the first row or column of that level. In general, the jth angle consists of cubes that are in the jth row or column but not in any previous angle. Each angle contains an odd number of cubes and the total number of cubes in each of the successive angles is strictly decreasing. Given a partition into strictly decreasing odd integers, this process is uniquely reversible. Each angle consists of at most $2r-1$ cubes; there are at most r angles per level; and there are at most t levels.

We now recompose our elements into a plane partition in which the ith stack of the jth column consists of those cubes that had been in the ith angle of the jth level (see Fig. 4.5). Since each level of the original plane partition does not overlap the level on which it sits, the new stack at position (i, j) is at least as high as the new stack at position $(i, j+1)$. We get the promised set of plane partitions inside $\mathcal{B}(r, t, 2r-1)$ for which every non-empty stack has odd height and a non-empty stack at position (i, j) must be strictly taller than the stack at position $(i+1, j)$.

Q.E.D.

We now pause to observe that, following exercise 4.2.8 on page 134, if we can prove that

$$\sum_{\lambda} s_{\lambda}(x_1,\ldots,x_n) = \prod_{i=1}^{n} \frac{1}{1-x_i} \prod_{1\le i<j\le n} \frac{1}{1-x_ix_j}, \tag{4.17}$$

then we have proven Gordon's result that the generating function for symmetric plane partitions with at most r rows or columns is given by

$$\prod_{i=1}^{r} \frac{1}{1-q^{2i-1}} \prod_{1\le i<j\le r} \frac{1}{1-q^{2i+2j-2}}.$$

While equation (4.17) *is* a special case of Lemma 4.5, we shall give a proof here because it illustrates the essential elements of the proof of Lemma 4.5 while avoiding many of the technical difficulties. Equation (4.17) was first proven by Issai Schur (1918).

Proof of equation (4.17)

We proceed by induction on the number of variables. When $n=1$, λ is restricted to be a partition with a single part, in which case $s_{\lambda_1}(x) = x^{\lambda_1}$. It follows that

$$\sum_{\lambda_1=0}^{\infty} s_{\lambda_1}(x) = \frac{1}{1-x}.$$

We assume that equation (4.17) is correct for fewer than n variables and we write

$$\begin{aligned} s_{\lambda}(x_1,\ldots,x_n) &= \frac{\det(x_j^{n+\lambda_i-i})}{\prod_{1\le i<j\le n}(x_i-x_j)} \\ &= \left(\prod_{1\le i<j\le n} \frac{1}{x_i-x_j}\right) \sum_{\sigma\in\mathcal{S}_n} (-1)^{\mathcal{I}(\sigma)} \prod_{i=1}^{n} x_{\sigma(i)}^{n+\lambda_i-i}. \end{aligned}$$

The quantity for which we want to find a compact formula is

$$\sum s_{\lambda} = \left(\prod_{1\le i<j\le n} \frac{1}{x_i-x_j}\right) \sum_{\lambda} \sum_{\sigma\in\mathcal{S}_n} (-1)^{\mathcal{I}(\sigma)} \prod_{i=1}^{n} x_{\sigma(i)}^{n+\lambda_i-i}.$$

We take this summation and first sum over the possible values of λ_n and $\sigma(n)$. If we subtract λ_n from each of the parts in λ, we are left with a partition, μ, into at most $n-1$ parts. We use Π to denote the Vandermonde product in front of the summation:

$$\begin{aligned}
\Pi &= \prod_{1\le i<j\le n}(x_i - x_j)^{-1},\\
\sum s_\lambda &= \Pi \sum_{\lambda_n=0}^{\infty}\sum_{k=1}^{n}\sum_{\mu}\sum_{\substack{\sigma\in\mathcal{S}_n\\ \sigma(n)=k}}(-1)^{\mathcal{I}(\sigma)}(x_1\cdots x_n)^{\lambda_n}\prod_{i=1}^{n-1}x_{\sigma(i)}^{n+\mu_i-i}\\
&= \Pi\,\frac{x_1\cdots x_n}{1-x_1\cdots x_n}\sum_{k=1}^{n}x_k^{-1}\sum_{\mu}\sum_{\substack{\sigma\in\mathcal{S}_n\\ \sigma(n)=k}}(-1)^{\mathcal{I}(\sigma)}\prod_{i=1}^{n-1}x_{\sigma(i)}^{n-1+\mu_i-i}.
\end{aligned}$$

We can now use our induction hypothesis on the sum over μ and permutations $\sigma\in\mathcal{S}_n$ for which $\sigma(n)=k$. In this summation, the $n-1$ variables exclude x_k. We do have to be careful about the inversion number. The difference between the inversion number of σ and the inversion number when k is removed is just the number of integers larger than and to the left of k in σ. Since k is the last letter of σ, this difference is $n-k$:

$$\begin{aligned}
\sum s_\lambda &= \Pi\,\frac{x_1\cdots x_n}{1-x_1\cdots x_n}\\
&\quad\times\sum_{k=1}^{n}x_k^{-1}(-1)^{n-k}\prod_{i\ne k}\frac{1}{1-x_i}\prod_{\substack{i<j\\ i,j\ne k}}\frac{1}{1-x_ix_j}\prod_{\substack{1\le i<j\le n\\ i,j\ne k}}(x_i-x_j)\\
&= \left(\prod_{i=1}^{n}\frac{1}{1-x_i}\prod_{1\le i<j\le n}\frac{1}{1-x_ix_j}\right)\frac{x_1\cdots x_n}{(1-x_1\cdots x_n)\prod_{i<j}(x_i-x_j)}\\
&\quad\times\sum_{k=1}^{n}(-1)^{n-k}x_k^{-1}(1-x_k)\prod_{i\ne k}(1-x_ix_k)\prod_{\substack{1\le i<j\le n\\ i,j\ne k}}(x_i-x_j).
\end{aligned}$$

We have reduced what we need to prove to the following identity,

$$\begin{aligned}
&(1-x_1\cdots x_n)\prod_{1\le i<j\le n}(x_i-x_j)\\
&\quad= x_1\cdots x_n\sum_{k=1}^{n}(-1)^{n-k}x_k^{-1}(1-x_k)\prod_{i\ne k}(1-x_ix_k)\prod_{\substack{1\le i<j\le n\\ i,j\ne k}}(x_i-x_j).
\end{aligned}\tag{4.18}$$

This can be proven by induction on n with the use of Proposition 2.8 (page 65). I leave it as an exercise to verify that equation (4.18) is correct for $n = 1$ or 2. I also leave it for you to check that the right side is an alternating polynomial in $x_1, \ldots, x_n$ (see exercise 4.3.2). If we divide the right side by $\prod(x_i - x_j)$, we get a symmetric polynomial in $x_1, \ldots, x_n$. Let us call this new polynomial $F_n(x_1, \ldots, x_n)$:

$$F_n(x_1, \ldots, x_n) = x_1 \cdots x_n \sum_{k=1}^{n} x_k^{-1}(1 - x_k) \prod_{i \neq k} \frac{1 - x_i x_k}{x_i - x_k}. \tag{4.19}$$

As a polynomial in x_1, F_n is a polynomial of degree n divided by a polynomial of degree $n-1$, and therefore it is a linear function of x_1. I leave it for you (exercise 4.3.4) to verify that this function is 1 when $x_1 = 0$ and, using the induction hypothesis, that it is equal to $1 - x_2 \cdots x_n$ when $x_1 = 1$. This implies that $F_n(x_1, \ldots, x_n) = 1 - x_1 \cdots x_n$.

Q.E.D.

An alternate form of equation (4.18) (whose equivalence is left as exercise 4.3.3) is

$$1 - x_1 \cdots x_n \;=\; x_1 \cdots x_n \sum_{k=1}^{n} x_k^{-1}(1 - x_k) \prod_{i \neq k} \frac{1 - x_i x_k}{x_i - x_k}. \tag{4.20}$$

Proof of Lemma 4.5

Our first step is to rewrite the lemma using the definition of the Schur function given in equation (4.3) and the $\mathbf{B_n}$ form of the Weyl denominator formula as given in equation (2.28). Note that in the definition of the Schur function, we set $n = r$ and take the transpose of the matrix. Equation (4.16) is equivalent to

$$\sum_{\lambda \subseteq \{t^r\}} \frac{\det(x_i^{\lambda_j + r - j})}{\prod_{1 \le i < j \le r}(x_i - x_j)}$$
$$= \frac{\det(x_i^{j-1} - x_i^{t+2r-j})}{\prod_{i=1}^{r}(1 - x_i) \prod_{1 \le i < j \le r}(x_i - x_j)(x_i x_j - 1)},$$

which, when we clear the denominators, becomes

$$\sum_{\lambda \subseteq \{t^r\}} \det(x_i^{\lambda_j+r-j}) \prod_{i=1}^{r}(1-x_i) \prod_{1 \le i < j \le r} (x_i x_j - 1) = \det(x_i^{j-1} - x_i^{t+2r-j}). \tag{4.21}$$

We shall prove equation (4.21) by induction on r, the number of variables. It is easy to see that it is true for $r = 1$. I leave it as an exercise for you to verify that it is correct for $r = 2$. Our induction hypothesis is that this equation is correct when we have at most $r - 1$ variables.

We write each determinant as a sum over permutations. For the determinant on the right, we must also specify a subset S of $\{1, 2, \dots, r\}$, the values of i for which we choose $-x_i^{t+2r-\sigma(i)}$ rather than $x_i^{\sigma(i)-1}$:

$$\begin{aligned} &\sum_{\lambda,\sigma}(-1)^{\mathcal{I}(\sigma)} \prod_{i=1}^{r} x_i^{\lambda_{\sigma(i)}+r-\sigma(i)} \prod_{i=1}^{r}(1-x_i) \prod_{1 \le i < j \le r} (x_i x_j - 1) \\ &\quad = \sum_{\sigma,S}(-1)^{\mathcal{I}(\sigma)+|S|} \prod_{i \in S} x_i^{t+2r-\sigma(i)} \prod_{i \notin S} x_i^{\sigma(i)-1} \end{aligned} \tag{4.22}$$

where the left-hand sum is over all partitions λ into at most r parts, each of which is less than or equal to t, and permutations σ, and the right-hand sum is over all permutations σ and subsets S of $\{1, 2, \dots, r\}$. This identity is equivalent to our lemma, and this is what we shall prove by induction. We assume that it is correct when we have fewer than r variables.

Let LHS denote the left-hand side of equation (4.22). For each partition $\lambda = (\lambda_1 \ge \cdots \ge \lambda_r \ge 0)$, there is a corresponding pair (μ, λ_r) where μ is the partition obtained from λ by subtracting λ_r from each of the first $r - 1$ parts, $\mu_i = \lambda_i - \lambda_r$, $1 \le i \le r - 1$. For each permutation $\sigma \in \mathcal{S}_r$, there is a corresponding pair (τ, k) where $k = \sigma^{-1}(r)$ and τ is a one-to-one mapping from $\{1, \dots, r\}\backslash\{k\}$ to $\{1, \dots, r-1\}$. If $\sigma^{-1}(r) = k$, then the sequence that corresponds to σ has an r in position k and therefore

$$\mathcal{I}(\sigma) = r - k + \mathcal{I}(\tau).$$

We can rewrite the left-hand side of equation (4.22) as

$$\begin{aligned} \text{LHS} &= \sum_{\lambda_r=0}^{t} \sum_{k=1}^{r} (-1)^{r-k}(1-x_k)x_k^{-1}(x_1 \cdots x_r)^{\lambda_r+1} \prod_{i \ne k}(x_i x_k - 1) \\ &\quad \times \sum_{\mu,\tau}(-1)^{\mathcal{I}(\tau)} \prod_{i \ne k} x_i^{\mu_{\tau(i)}+(r-1)-\tau(i)} \prod_{\substack{i=1 \\ i \ne k}}^{r}(1-x_i) \prod_{\substack{1 \le i < j \le r \\ i,j \ne k}} (x_i x_j - 1). \end{aligned}$$

We now apply our induction hypothesis to the sum on μ and τ:

$$\begin{aligned}\text{LHS} &= \sum_{\lambda_r=0}^{t}\sum_{k=1}^{r}(-1)^{r-k}(1-x_k)x_k^{-1}(x_1\cdots x_r)^{\lambda_r+1}\prod_{i\neq k}(x_i\,x_k-1)\\ &\qquad\times\sum_{\sigma,S}(-1)^{\mathcal{I}(\sigma)+|S|}\prod_{i\in S}x_i^{t-\lambda_r+2(r-1)-\sigma(i)}\prod_{i\in\overline{S}}x_i^{\sigma(i)-1}\\ &= \sum_{\lambda_r=0}^{t}\sum_{k=1}^{r}\sum_{\sigma}\sum_{S}(-1)^{r-k+\mathcal{I}(\sigma)+|S|}(1-x_k)x_k^{-1}\prod_{i\neq k}(x_ix_k-1)\\ &\qquad\times\prod_{i\notin S}x_i^{\lambda_r+1}\prod_{i\in S}x_i^{t+2r-2}\prod_{i\in S}x_i^{1-\sigma(i)}\prod_{i\in\overline{S}}x_i^{\sigma(i)-1}.\end{aligned}$$

where S is a subset of $\{1,\ldots,r\}\backslash\{k\}$, $\overline{S}$ is the complement of S in $\{1,\ldots,r\}\backslash\{k\}$, and σ is a one-to-one mapping from $\{1,\ldots,r\}\backslash\{k\}$ to $\{1,\ldots,r-1\}$.

We can simplify the summation on λ_r:

$$\sum_{\lambda_r=0}^{t}\prod_{i\notin S}x_i^{\lambda_r}=\frac{1-\prod_{i\notin S}x_i^{t+1}}{1-\prod_{i\notin S}x_i}.$$

We can also simplify the summation on σ using the Vandermonde identity:

$$\sum_{\sigma}(-1)^{\mathcal{I}(\sigma)}\prod_{i\in S}x_i^{1-\sigma(i)}\prod_{i\in\overline{S}}x_i^{\sigma(i)-1}=(-1)^{\binom{r-1}{2}}\prod_{\substack{1\le i<j\le r\\ i,j\neq k}}(x_i^{\epsilon_i}-x_j^{\epsilon_j}),$$

where

$$\epsilon_i=\begin{cases}-1, & i\in S,\\ +1, & i\notin S.\end{cases}$$

This means that we can write the left-hand side of equation (4.22) as

$$\begin{aligned}\text{LHS} &= (-1)^{\binom{r-1}{2}}\sum_{k=1}^{r}\sum_{S\subseteq\{1,\ldots,r\}\backslash\{k\}}(-1)^{|S|+r-k}(1-x_k)x_k^{-1}\\ &\qquad\times\prod_{i\notin S}x_i\left(\frac{1-\prod_{i\notin S}x_i^{t+1}}{1-\prod_{i\notin S}x_i}\right)\\ &\qquad\times\prod_{i\neq k}(x_ix_k-1)\prod_{i\in S}x_i^{t+2r-2}\prod_{\substack{1\le i<j\le r\\ i,j\neq k}}(x_i^{\epsilon_i}-x_j^{\epsilon_j}).\end{aligned}$$

We can write

$$\begin{aligned}\prod_{i\neq k}(x_ix_k-1) &= (-1)^{k-1}\prod_{\substack{i<k\\ i\notin S}}(1-x_ix_k)\prod_{\substack{k<i\\ i\notin S}}(x_ix_k-1)\\ &\quad\times\prod_{i\in S}x_i\prod_{\substack{i<k\\ i\in S}}(x_i^{-1}-x_k)\prod_{\substack{i>k\\ i\in S}}(x_k-x_i^{-1}).\end{aligned}$$

We use this equality to rewrite

$$\begin{aligned}&\prod_{i\neq k}(x_ix_k-1)\prod_{\substack{1\le i<j\le r\\ i,j\neq k}}(x_i^{\epsilon_i}-x_j^{\epsilon_j})\\ &\quad= \prod_{\substack{i<j\\ i\text{ or } j\in S}}(x_i^{\epsilon_i}-x_j^{\epsilon_j})\prod_{\substack{i<j\\ i,j\neq k,\ \notin S}}(x_i-x_j)\\ &\qquad\times(-1)^{k-1}\prod_{i\in S}x_i\prod_{i<k}(1-x_ix_k)\prod_{k<i}(x_ix_k-1).\end{aligned}$$

We make this substitution and then interchange the summation on S and k. The sum on S becomes a sum over all *proper* subsets of $\{1,\ldots,r\}$. We get that

$$\begin{aligned}\text{LHS} &= (-1)^{\binom{r}{2}}\sum_{S\subset\{1,\ldots,r\}}(-1)^{|S|}\prod_{i\in S}x_i^{t+2r-1}\\ &\quad\times\prod_{\substack{1\le i<j\le r\\ i\text{ or } j\in S}}(x_i^{\epsilon_i}-x_j^{\epsilon_j})\left(\frac{1-\prod_{i\notin S}x_i^{t+1}}{1-\prod_{i\notin S}x_i}\right)\\ &\quad\times\prod_{i\notin S}x_i\sum_{k\notin S}(1-x_k)x_k^{-1}\prod_{\substack{i<k\\ i\notin S}}(1-x_i\,x_k)\\ &\quad\times\prod_{\substack{i>k\\ i\notin S}}(x_i\,x_k-1)\prod_{\substack{1\le i<j\le r\\ i,j\notin S,\ \neq k}}(x_i-x_j).\end{aligned}$$

We can write these last two lines as

$$\prod_{\substack{1\le i<j\le r\\ i,j\notin S}}(x_i-x_j)\prod_{i\notin S}x_i\sum_{k\notin S}(1-x_k)x_k^{-1}\prod_{i\neq k,\ \notin S}\frac{1-x_ix_k}{x_i-x_k}.$$

By equation (4.20), this is equal to

$$(1-\prod_{i\notin S}x_i)\prod_{\substack{i<j\\ i,j\notin S}}(x_i-x_j).$$

We make this replacement and combine the Vandermonde products:

$$\begin{aligned}
\text{LHS} &= (-1)^{\binom{r}{2}} \sum_{S\subset\{1,\ldots,r\}} (-1)^{|S|} \prod_{i\in S} x_i^{t+2r-1} \prod_{\substack{1\le i<j\le r \\ i \text{ or } j\in S}} (x_i^{\epsilon_i} - x_j^{\epsilon_j}) \\
&\quad \times \left(1 - \prod_{i\notin S} x_i^{t+1}\right) \prod_{\substack{1\le i<j\le r \\ i,j\notin S}} (x_i - x_j) \\
&= (-1)^{\binom{r}{2}} \left(\sum_{S\subset\{1,\ldots,r\}} (-1)^{|S|} \prod_{i\in S} x_i^{t+2r-1} \prod_{1\le i<j\le n} (x_i^{\epsilon_i} - x_j^{\epsilon_j}) \right. \\
&\quad \left. - \sum_{S\subset\{1,\ldots,r\}} (-1)^{|S|} \prod_{i\in S} x_i^{t+2r-1} \prod_{i\notin S} x_i^{t+1} \prod_{1\le i<j\le r} (x_i^{\epsilon_i} - x_j^{\epsilon_j}) \right).
\end{aligned}$$

We expand our Vandermonde product:

$$\begin{aligned}
\text{LHS} = &\sum_{\sigma,\ S\subset\{1,\ldots,r\}} (-1)^{\mathcal{I}(\sigma)+|S|} \prod_{i\in S} x_i^{t+2r-\sigma(i)} \prod_{i\notin S} x_i^{\sigma(i)-1} \\
&- \sum_{\sigma,\ S\subset\{1,\ldots,r\}} (-1)^{\mathcal{I}(\sigma)+|S|} \prod_{i\in S} x_i^{t+2r-\sigma(i)} \prod_{i\notin S} x_i^{t+\sigma(i)}.
\end{aligned} \tag{4.23}$$

This looks very much like the right-hand side of equation (4.22). The only difference is that we are missing the terms where $S = \{1, \ldots, r\}$ and we are subtracting the second summation. As we now shall see, the second summation gives us precisely the missing summand.

We observe that

$$\det\left(x_i^{t+j} - x_i^{t+2r-j}\right)_{i,j=1}^{r} = 0.$$

This is because the rth column of this matrix consists of nothing but 0s. If we rewrite this determinant as a sum over permutations and subsets, we see that

$$\begin{aligned}
&\sum_{\sigma} (-1)^{\mathcal{I}(\sigma)+r} \prod_{i} x_i^{t+2r-\sigma(i)} \\
&\qquad = - \sum_{\sigma,\ S\subset\{1,\ldots,r\}} (-1)^{\mathcal{I}(\sigma)+|S|} \prod_{i\in S} x_i^{t+2r-\sigma(i)} \prod_{i\notin S} x_i^{t+\sigma(i)}.
\end{aligned}$$

The term on the left of this equality is precisely the $S = \{1, \ldots, r\}$ term of the first summation on the right-hand side of equation (4.23). We make this substitution:

$$\text{LHS} = \sum_{\sigma,\ S\subseteq\{1,\ldots,r\}} (-1)^{\mathcal{I}(\sigma)+|S|} \prod_{i\in S} x_i^{t+2r-\sigma(i)} \prod_{i\notin S} x_i^{\sigma(i)-1}.$$

Q.E.D.

Conclusion of the proof of MacMahon's conjecture

When we put together Lemmas 4.4 and 4.5, we see that the generating function for symmetric plane partitions that fit inside $\mathcal{B}(r,r,t)$ is

$$\sum_{\lambda\subseteq\{t^r\}} s_\lambda(q^{2r-1}, q^{2r-3}, \ldots, q) = X/Y$$

where

$$\begin{aligned} X &= \det\left((q^{2r-2i+1})^{j-1} - (q^{2r-2i+1})^{t+2r-j}\right), \\ Y &= \det\left((q^{2r-2i+1})^{j-1} - (q^{2r-2i+1})^{2r-j}\right). \end{aligned}$$

We use the $\mathbf{B_n}$ form of the Weyl denominator formula, equation (2.28), to express each of these determinants as a product. If we set $x_i = q^{2r-2i+1}$ in the Weyl denominator formula and then do some simplification, we see that

$$Y = \prod_{i=1}^{n}(1-q^{2i-1}) \prod_{1\leq i<j\leq n} \left[(1-q^{2i+2j-2})(1-q^{2j-2i})q^{2i-1}\right]. \quad (4.24)$$

The numerator will require a bit more ingenuity. We first rewrite the determinant in the $\mathbf{B_n}$ form of the Weyl denominator formula as

$$\begin{aligned} &\det(x_j^{i-1} - x_j^{2r-i}) \\ &\quad = \prod_{i=1}^{r} x_i^{(2r-1)/2} \sum_{\substack{\sigma\in S_r \\ S\subseteq\{1,\ldots,r\}}} (-1)^{\mathcal{I}(\sigma)+|S|} \prod_{i=1}^{r} x_{\sigma(i)}^{\epsilon_i(2i-1-2r)/2}, \quad (4.25) \end{aligned}$$

where, as before, $\epsilon_i = -1$ if $i \in S$ and $\epsilon_i = +1$ if $i \notin S$. Using this same idea, we can write

$$\begin{aligned} &\det(x_j^{i-1} - x_j^{t+2r-i}) \\ &\quad = \prod_{i=1}^{r} x_i^{(t+2r-1)/2} \sum_{\substack{\sigma\in S_r \\ S\subseteq\{1,\ldots,r\}}} (-1)^{\mathcal{I}(\sigma)+|S|} \prod_{i=1}^{r} x_{\sigma(i)}^{\epsilon_i(2i-1-2r-t)/2}. \end{aligned}$$

For the double sum on σ and S, we can replace σ by σ^{-1} – this does not change the inversion number – and each ϵ_i by $\epsilon_{\sigma^{-1}(i)}$ – this does not change the cardinality of S. Since the double sum is over all possible pairs, (σ, S), we can then replace i by $\sigma(i)$ in the product:

$$\sum_{\substack{\sigma\in\mathcal{S}_r \\ S\subseteq\{1,\ldots,r\}}} (-1)^{\mathcal{I}(\sigma)+|S|} \prod_{i=1}^{r} x_{\sigma(i)}^{\epsilon_i(2i-1-2r-t)/2}$$
$$= \sum_{\substack{\sigma\in\mathcal{S}_r \\ S\subseteq\{1,\ldots,r\}}} (-1)^{\mathcal{I}(\sigma)+|S|} \prod_{i=1}^{r} x_i^{\epsilon_i(2\sigma(i)-1-2r-t)/2}.$$

We now pull these observations together,

$$X = \prod_{i=1}^{r} (q^{2r-2i+1})^{(t+2r-1)/2}$$
$$\times \sum_{\substack{\sigma\in\mathcal{S}_r \\ S\subseteq\{1,\ldots,r\}}} (-1)^{\mathcal{I}(\sigma)+|S|} \prod_{i=1}^{r} q^{\epsilon_i(t+2r-2\sigma(i)+1)(2i-1-2r)/2},$$

and then use equation (4.25) and simplify the power of q,

$$X = q^{-tr(r-1)/2} \det\left((q^{t+2r-2j+1})^{i-1} - (q^{t+2r-2j+1})^{2r-i}\right).$$

We can now apply the $\mathbf{B_n}$ form of the Weyl denominator formula to write the determinant in our numerator as a product, and then simplify to get

$$X = \prod_{i=1}^{r}(1-q^{t+2i-1}) \prod_{1\le i<j\le r} \left[(1-q^{2t+2i+2j-2})(1-q^{2j-2i})q^{2i-1}\right]. \tag{4.26}$$

We can use equations (4.24) and (4.26) to write our generating function as a product:

$$\sum_{\lambda\subseteq\{t^r\}} s_\lambda(q^{2r-1}, q^{2r-3}, \ldots, q)$$
$$= \frac{\prod_{i=1}^{r}(1-q^{t+2i-1}) \prod_{1\le i<j\le r} \left[(1-q^{2t+2i+2j-2})(1-q^{2j-2i})q^{2i-1}\right]}{\prod_{i=1}^{r}(1-q^{2i-1}) \prod_{1\le i<j\le r} \left[(1-q^{2i+2j-2})(1-q^{2j-2i})q^{2i-1}\right]}$$
$$= \prod_{i=1}^{r} \frac{1-q^{t+2i-1}}{1-q^{2i-1}} \prod_{1\le i<j\le r} \frac{1-q^{2t+2i+2j-2}}{1-q^{2i+2j-2}},$$

which is the formula, (4.14), that we set out to prove.

Q.E.D.

We have scaled the first of our summits, the proof of Conjecture 4. As we saw in Chapter 1, the structure of this generating function led Macdonald to Conjecture 6, the generating function for cyclically symmetric plane partitions.

Part of the interest in Macdonald's conjecture was the belief that its proof might also rely on new insights into symmetric functions. As we shall see, the proof of Conjecture 6 requires a very different set of tools and yields very different insights.

Exercises

4.3.1 Verify that equation (4.18) is correct for $n = 1$ and $n = 2$.

4.3.2 Prove that the right side of equation (4.18) is an alternating polynomial in $x_1, \ldots, x_n$.

4.3.3 Prove that equation (4.18) implies equation (4.20).

4.3.4 Prove that the right side of equation (4.19) is 1 when $x_1 = 0$. Use the induction hypothesis to prove that the right side of equation (4.19) is $1 - x_2 \cdots x_n$ when $x_1 = 1$.

4.3.5 Verify that $F_n(0, x_2, \ldots, x_n) = 0$ and (using the induction hypothesis) that $F_n(1, x_2, \ldots, x_n) = 1 - x_2 \cdots x_n$.

4.3.6 Verify that Lemma 4.5 is correct for $n = 1$ and $n = 2$.

4.3.7 Use Lemma 4.5 to prove that

$$\sum_{\lambda \subseteq t^r} s_\lambda(q, q^2, \ldots, q^r) = \prod_{1 \le i \le j \le r} \frac{1 - q^{t+i+j-1}}{1 - q^{i+j-1}}, \tag{4.27}$$

where the sum is over all partitions λ with at most r parts, each of which is less than or equal to t. This implies that the generating function for column strict plane partitions with at most r rows, t columns, and largest part at most r is given by

$$\prod_{1 \le i \le j \le r} \frac{1 - q^{t+i+j-1}}{1 - q^{i+j-1}},$$

a result that is known as the Bender–Knuth conjecture (Bender and Knuth 1972).

4.3.8 Prove that

$$x_1 \cdots x_n \sum_{k=1}^{n} x_k^{-1}(1 - t x_k)(1 - v x_k) \prod_{\substack{i=1 \\ i \ne k}}^{n} \frac{1 - x_i x_k}{x_i - x_k}$$

$$= \begin{cases} (1 - t x_1 \cdots x_n)(1 - v x_1 \cdots x_n), & \text{if } n \text{ is odd,} \\ (1 - x_1 \cdots x_n)(1 - t v x_1 \cdots x_n), & \text{if } n \text{ is even.} \end{cases} \tag{4.28}$$

4.3.9 Use equation (4.28) to prove that

$$\sum_{\lambda} f_\lambda(t,v) s_\lambda(x_1,\ldots,x_n) = \prod_{i=1}^{n} \frac{1}{(1-tx_i)(1-vx_i)} \prod_{1\le i<j\le n} \frac{1}{1-x_i x_j}, \tag{4.29}$$

where we let a_j be the number of columns of length j in λ (equivalently, the number of parts of size j in λ') and then

$$f_\lambda(t,v) = \prod_{j \text{ odd}} \frac{v^{a_j+1} - t^{a_j+1}}{v-t} \prod_{j \text{ even}} \frac{1-(tv)^{a)j+1}}{1-tv}.$$

4.3.10 With f_λ defined in exercise 4.3.9, prove that

$$\begin{aligned} f_\lambda(0,1) &= 1, \\ f_\lambda(1,-1) &= \begin{cases} 0 & \text{if any } a_j \text{ is odd,} \\ 1 & \text{otherwise,} \end{cases} \\ f_\lambda(0,0) &= \begin{cases} 0 & \text{if any } a_j \text{ is positive for any odd } j, \\ 1 & \text{otherwise.} \end{cases} \end{aligned}$$

Show that the following are special cases of equation (4.29):

$$\begin{aligned} \sum_{\lambda} s_\lambda(x_1,\ldots,x_n) &= \prod_{i=1}^{n} \frac{1}{1-x_i} \prod_{1\le i<j\le n} \frac{1}{1-x_i x_j}, \\ \sum_{\lambda \text{ even}} s_\lambda(x_1,\ldots,x_n) &= \prod_{i=1}^{n} \frac{1}{1-x_i^2} \prod_{1\le i<j\le n} \frac{1}{1-x_i x_j}, \\ \sum_{\lambda' \text{ even}} s_\lambda(x_1,\ldots,x_n) &= \prod_{1\le i<j\le n} \frac{1}{1-x_i x_j}. \end{aligned}$$

The last two identities are due to D. E. Littlewood (1950).

4.3.11 Prove that

$$x_1 \cdots x_n \sum_{k=1}^{n} x_k^{-1}(1-t\,x_k) \prod_{i\ne k} \frac{1-v\,x_i x_k}{x_i - x_k} = 1 - t^{\chi(n \text{ odd})} v^{\lfloor n/2 \rfloor} x_1 \cdots x_n, \tag{4.30}$$

where $\chi(S) = 1$ if S is true and 0 if S is false.

4.3.12 Let $c(\lambda)$ be the number of odd parts in the partition conjugate to λ and let $d(\lambda) = \sum_{i\ge 1} \nu_i \lfloor i/2 \rfloor$ where $\nu_i = \nu_i(\lambda)$ is the number parts of size i in the partition conjugate to λ. Use equation (4.30) to prove that

$$\sum_{\lambda} t^{c(\lambda)} v^{d(\lambda)} s_\lambda(x_1, \ldots, x_n) = \prod_{i=1}^{n} \frac{1}{1 - t\, x_i} \prod_{1 \le i < j \le n} \frac{1}{1 - v\, x_i x_j}. \tag{4.31}$$

4.3.13 Prove that the parity of $d(\lambda)$ is the same as the parity of $n(\lambda) = \sum_{i \ge 1} (i-1)\lambda_i$, and therefore

$$\sum_{\lambda} (-1)^{n(\lambda)} s_\lambda(x_1, \ldots, x_n) = \prod_{i=1}^{n} \frac{1}{1 - x_i} \prod_{1 \le i < j \le n} \frac{1}{1 + x_i x_j}. \tag{4.32}$$

One of the most important sets of mathematics books written in the last twenty years is Knuth's *The Art of Computer Programming* (1968, 1969, 1973). In this set of books, and in many others of a combinatorial nature, binomial coefficients occur regularly.... Knuth wrote:

"There are literally thousands of identities involving binomial coefficients, and for centuries many people have been pleased to discover them. However, there are so many relations present that when someone finds a new identity, there aren't many people who get excited about it any more, except the discoverer! In order to manipulate the formulas which arise in the analysis of algorithms, a facility for handling binomial coefficients is a must." (1968, sect. 1.2.6, 52–53)

When a mathematician who is as good as Knuth writes nonsense like the above (except for the last sentence, where he is probably right), then one must look seriously at what he wrote and try to understand why he missed the essence of what is really true. There are actually very few identities of the sort Knuth gave in this section – there just seem to be many because he does not know how to write them. For example, in (1968, sect. 1.2.6I) he gave six sums (21)–(26) and then wrote that (21) is by far the most important. What he did not point out is that five of these identities, (21)–(25), are all just disguised versions of

$$ {}_2F_1\left[\begin{matrix} -n, & a \\ & c \end{matrix} ;1\right] = \frac{(c-a)_n}{(c)_n}. $$

In other words, they are all the same identity. Binomial coefficients are important, since they count things; but when one has a series of products of binomial coefficients, the right thing to do is to translate the sum to the hypergeometric series.... Translation is almost always easy (there can be some problems that require limits when division by zero arises), and it has been known for a long time that this is the right way to handle sums of products of binomial coefficients. Andrews spelled this out in detail in (1974, sect. 5), but the realization that hypergeometric series are just series with term ratio a rational function of n [the index of summation] is very old. Horn (1889) used this as the definition of a hypergeometric series in two variables. R. Narasimhan told me that he found a definition of "comfortable" series in one of the late volumes of Euler's collected works. For Euler, a comfortable series is a power series whose term ratio is a rational function of n. When I asked Narasimhan to give me a specific reference, he was unable to find it again. I will be very pleased to pay \$50 U.S.† for this reference, for it would be worth that to know that Euler's insight was also good here. An even earlier place one might look for this insight would be in Newton's work.

– Richard Askey, from How can mathematicians and mathematical historians help each other? (1988, 206–207)

† Professor Askey informs me that this offer still stands.

5
Hypergeometric Series

At a lecture on hypergeometric functions I attended at the University of Edinburgh, the speaker, after foisting on us a colossal equation, announced grandly, "This result has applications in particle physics." He then added, roguishly, "Of course, *every* result has applications in particle physics – the field is in such disarray." I seemed to be the only one in the audience who was amused.

– Jet Wimp (1997)

Macdonald had used results from the theory of representations of the symmetric group to prove MacMahon's conjecture. At the same time, George Andrews found a proof that drew on an entirely different tradition, the special functions of analysis that can be expressed as hypergeometric series. Connections between analysis and combinatorics go back at least to Euler, but the collaborations of Richard Askey and George Andrews, beginning in the 1970s, revealed powerful new links between partition theory and special functions.

Hypergeometric series play a small role in the Mills, Robbins, and Rumsey proof of the Macdonald conjecture. Their proof is really a tour de force of the techniques of linear algebra. But as algebraic combinatorics developed in the 1980s, the machinery of hypergeometric series became indispensable to its study. The proof of the refined alternating sign matrix conjecture will rely on some sophisticated pieces of this machinery.

5.1 Mills, Robbins, and Rumsey's bright idea

The beauty of the proof of Macdonald's conjecture for the number of cyclically symmetric plane partitions lies in its structure and its use of inspired guesswork and serendipitous happenstance. Almost all of this can be found in the simplest case, a proof of Andrews's result that counts the total number of descending plane partitions with largest part less than or equal to r. This is the $q = 1$ case of Andrews's conjecture (Conjecture 7). In the first section, I shall outline this proof and show how to proceed in general. I shall present the details and elaborations needed for the full proof of the Macdonald conjecture at the end of this chapter, after deriving the results we need from the theory of hypergeometric series.

For Mills, Robbins, and Rumsey, the starting point was the same as it was for Andrews, the determinant that expresses the number to be calculated. We define I_r to be the $r \times r$ identity matrix. The matrix H_r is given by

$$H_r := \left(\binom{i+j}{j-1} \right)_{1 \le i,j \le r}.$$

As we saw in Chapter 3, Theorem 3.11, the number of descending plane partitions with largest part less than or equal to $r+1$ is the determinant of $I_r + H_r$. Our task is to prove that for all positive integers r,

$$\det(I_r + H_r) = \prod_{j=0}^{r} \frac{(3j+1)!}{(r+j+1)!}.$$

As an example, when $r = 5$ we are counting the number of descending plane partitions with largest part less than or equal to 6. This number is

$$\det(I_5 + H_5) = \begin{vmatrix} 1+1 & 3 & 6 & 10 & 15 \\ 1 & 1+4 & 10 & 20 & 35 \\ 1 & 5 & 1+15 & 35 & 70 \\ 1 & 6 & 21 & 1+56 & 126 \\ 1 & 7 & 28 & 84 & 1+210 \end{vmatrix} = 7436.$$

Decomposing the determinant

The insight drawn from the correspondence with alternating sign matrices suggests that we want to investigate the numbers $L_i = L_{i,r} =$ the number of descending plane partitions with largest part less than or equal to $r + 1$ and exactly i parts of size $r + 1$. In Theorem 3.11, page 108, we proved that

$$L_i = L_{i,r} = \det(H_{i,r}),$$

where $H_{i,r}$ is identical to $I_r + H_r$ in the first $r-1$ rows. The rth row of $H_{i,r}$ is

$$\begin{array}{cl} (0,0,\ldots,0,1), & \text{if } i=0, \\ \left(\binom{r+1-i}{1-i},\binom{r+2-i}{2-i},\ldots,\binom{2r-i}{r-i}\right), & \text{if } 1 \le i \le r. \end{array}$$

The binomial coefficient is taken to be 0 when the lower parameter is a negative integer:

$$\binom{a}{-n} = 0 \qquad \text{for } n \in \mathbf{N}.$$

For $r=5$, the values of the L_i are

$$L_0 = L_5 = \begin{vmatrix} 2 & 3 & 6 & 10 & 15 \\ 1 & 5 & 10 & 20 & 35 \\ 1 & 5 & 16 & 35 & 70 \\ 1 & 6 & 21 & 57 & 126 \\ 0 & 0 & 0 & 0 & 1 \end{vmatrix} = 429,$$

$$L_1 = \begin{vmatrix} 2 & 3 & 6 & 10 & 15 \\ 1 & 5 & 10 & 20 & 35 \\ 1 & 5 & 16 & 35 & 70 \\ 1 & 6 & 21 & 57 & 126 \\ 1 & 6 & 21 & 56 & 126 \end{vmatrix} = 1287,$$

$$L_2 = \begin{vmatrix} 2 & 3 & 6 & 10 & 15 \\ 1 & 5 & 10 & 20 & 35 \\ 1 & 5 & 16 & 35 & 70 \\ 1 & 6 & 21 & 57 & 126 \\ 0 & 1 & 6 & 21 & 56 \end{vmatrix} = 2002,$$

$$L_3 = \begin{vmatrix} 2 & 3 & 6 & 10 & 15 \\ 1 & 5 & 10 & 20 & 35 \\ 1 & 5 & 16 & 35 & 70 \\ 1 & 6 & 21 & 57 & 126 \\ 0 & 0 & 1 & 6 & 21 \end{vmatrix} = 2002,$$

$$L_4 = \begin{vmatrix} 2 & 3 & 6 & 10 & 15 \\ 1 & 5 & 10 & 20 & 35 \\ 1 & 5 & 16 & 35 & 70 \\ 1 & 6 & 21 & 57 & 126 \\ 0 & 0 & 0 & 1 & 6 \end{vmatrix} = 1287.$$

Conjecture 9 is equivalent to the statement that

$$L_i = \binom{r+i}{i} \frac{(2r-i)!}{(r-i)!} \prod_{j=0}^{r-1} \frac{(3j+1)!}{(r+j+1)!}.$$

The sum of the last rows of these $r+1$ matrices is the last row of $I_r + H_r$ (I leave this as an exercise). It follows that

$$\det(I_r + H_r) = \sum_{i=0}^{r} L_i.$$

This will be a proof by induction on r. We may assume that $\det(I_{r-1} + H_{r-1})$ is the predicted value:

$$L_0 = \det(I_{r-1} + H_{r-1}) = \prod_{j=0}^{r-1} \frac{(3j+1)!}{(r+j)!}.$$

More definitions

We let H_r^* be the matrix H_r with the last row replaced by a row of 0s so that

$$H_{0,r} = I_r + H_r^*.$$

We let $R = R_r$ denote the matrix whose ith row is the last row of $H_{i,r}$. For example, when $r = 5$ we have

$$R_5 = \begin{pmatrix} 1 & 6 & 21 & 56 & 126 \\ 0 & 1 & 6 & 21 & 56 \\ 0 & 0 & 1 & 6 & 21 \\ 0 & 0 & 0 & 1 & 6 \\ 0 & 0 & 0 & 0 & 1 \end{pmatrix}.$$

The matrix R is always upper triangular with 1s on the diagonal, and therefore it is always nonsingular. Finally, we let $C = C_r$ denote the column vector of cofactors† of the last row of $I_r + H_r$. Since each matrix $H_{i,r}$ only differs from $I_r + H_r$ in the last row, the matrix product of the last row of $H_{i,r}$ with C is the determinant of $H_{i,r}$. This determinant is

† The cofactor of entry i, j in matrix M is $(-1)^{i+j}$ times the determinant of the matrix obtained from M by deleting row i and column j. This becomes entry j, i in the **cofactor matrix**. Our column vector C is the rightmost column of the cofactor matrix.

L_i. The matrix product of any other row of $H_{i,r}$ with C is the determinant of the matrix obtained by replacing the last row of $H_{i,r}$ with this other row of the same matrix. This determinant is zero. We have that

$$H_{i,r}C = \begin{pmatrix} 0 \\ 0 \\ \vdots \\ 0 \\ L_i \end{pmatrix}.$$

Because R is the matrix of these last rows, we also have that

$$RC = \begin{pmatrix} L_1 \\ L_2 \\ \vdots \\ L_r \end{pmatrix}.$$

Since R is nonsingular, this can be rewritten as

$$R^{-1}\begin{pmatrix} L_1 \\ L_2 \\ \vdots \\ L_r \end{pmatrix} = C. \tag{5.1}$$

An equation that uniquely determines the L_i

Using equation (5.1) and the fact that $I_r + H_r^* = H_{0,r}$, we see that

$$\begin{aligned} R(I_r + H_r^*)R^{-1}\begin{pmatrix} L_1 \\ L_2 \\ \vdots \\ L_m \end{pmatrix} &= R(I_r + H_r^*)C \\ &= RH_{0,r}C. \end{aligned}$$

The vector C is the column vector of cofactors of the last row of $I_r + H_r$. This implies that if we take any row of $H_{0,r}$ other than the last row and find its dot product with C, the result will be 0. The dot product of the last row of $H_{0,r}$ with C is the determinant of $H_{0,r}$ which is the scalar

L_0. We can rewrite our equation as

$$R(I_r + H_r^*)R^{-1} \begin{pmatrix} L_1 \\ L_2 \\ \vdots \\ L_r \end{pmatrix} = R \begin{pmatrix} 0 \\ 0 \\ \vdots \\ 0 \\ L_0 \end{pmatrix}$$

$$= L_0 R \begin{pmatrix} 0 \\ 0 \\ \vdots \\ 0 \\ 1 \end{pmatrix}.$$

When we multiply R by this column vector, we simply pick off the last column of R,

$$R(I_r + H_r^*)R^{-1} \begin{pmatrix} L_1 \\ L_2 \\ \vdots \\ L_r \end{pmatrix} = L_0 \begin{pmatrix} R_{1,r} \\ R_{2,r} \\ \vdots \\ R_{r,r} \end{pmatrix}, \tag{5.2}$$

where $R_{i,r}$ is the entry in row i, column r of R,

$$R_{i,r} = \binom{2r-i}{r-i}.$$

The determinant of $R(I_r + H_r^*)R^{-1}$ is the determinant of $I_r + H_r^* = H_{0,r}$ which is L_0. We know the value of L_0 (by the induction hypothesis), and it is not zero. Therefore, $R(I_r + H_r^*)R^{-1}$ is nonsingular. In particular, the values of L_1 through L_r are uniquely determined by equation (5.2). If we can show that our conjectured product formulæ for the values of the L_i satisfy this equation, then we have proven that they are correct.

For the problem at hand, that of verifying the formula for the total number of descending plane partitions with largest part less than or equal to $r+1$ and exactly i parts of size $r+1$, we are done. The prospect may be daunting, but it is possible to show that the conjectured values satisfy this equation and to do so using only the tools we now have on hand. We run into a problem, however, when we try to verify the product form of the generating function for descending plane partitions or to prove the Macdonald conjecture. The values of L_i are now polynomials in q, and we do not know what they should be. We could guess and hope to be lucky. Mills, Robbins, and Rumsey took a more systematic approach.

The leap of faith

Equation (5.2) is equivalent to

$$\begin{pmatrix} L_1 \\ L_2 \\ \vdots \\ L_r \end{pmatrix} = L_0 \begin{pmatrix} R_{1,r} \\ R_{2,r} \\ \vdots \\ R_{r,r} \end{pmatrix} - RH_r^* R^{-1} \begin{pmatrix} L_1 \\ L_2 \\ \vdots \\ L_r \end{pmatrix}.$$

This matrix equation represents r linear equations in which the right-hand side of each is a sum of $r+1$ terms. We can rewrite this right-hand side as a single matrix product:

$$\begin{pmatrix} L_1 \\ L_2 \\ \vdots \\ L_r \end{pmatrix} = \left(\begin{array}{c|c} R_{1,r} & \\ R_{2,r} & \\ \vdots & -RH_r^* R^{-1} \\ R_{r,r} & \end{array} \right) \begin{pmatrix} L_0 \\ L_1 \\ L_2 \\ \vdots \\ L_r \end{pmatrix}. \tag{5.3}$$

This identity expresses a column vector of length r as the product of an $r \times (r+1)$ matrix and a column vector of length $r+1$. Following Mills, Robbins, and Rumsey, we consider extending this to a system of $r+1$ equations, a consideration that begins to seem reasonable when one calculates the entries of the matrix and discovers that

$$\left(\begin{array}{c|c} R_{1,r} & \\ R_{2,r} & \\ \vdots & -RH_r^* R^{-1} \\ R_{r,r} & \end{array} \right) = \left((-1)^j \binom{2r-i}{r-i-j} \right)_{\substack{1 \le i \le r \\ 0 \le j \le r}}.$$

Equation (5.3) strongly suggests that we should define an $(r+1) \times (r+1)$ matrix

$$K = K_r := \left((-1)^j \binom{2r-i}{r-i-j} \right)_{i,j=0}^{r}$$

and expect that

$$\begin{pmatrix} L_0 \\ L_1 \\ L_2 \\ \vdots \\ L_r \end{pmatrix} = K \begin{pmatrix} L_0 \\ L_1 \\ L_2 \\ \vdots \\ L_r \end{pmatrix}. \tag{5.4}$$

Let us pause for a moment to appreciate the audacity of what we are proposing. We know that equation (5.3) is correct and that if our product formulæ for the L_i satisfy this equation, then they have been proven correct. Equation (5.4) implies equation (5.3) – we have simply added an extra equation in a system of equations – and therefore if our product formulæ satisfy equation (5.4) then they have been proven to be correct. But remember that, by our induction hypothesis, L_0 is a known quantity. Equation (5.4) is a system of $r+1$ equations in r unknowns. There is no guarantee that it has a solution. The only reason to proceed is the observation that we seem to have completed a missing symmetry. To a mathematician, there could be no better reason.

Equation (5.4) is very appealing for it states that $L = (L_0, L_1, \ldots, L_r)$ is an eigenvector of K with eigenvalue 1. In order to complete the inductive step of our proof we need only find some eigenvector for the eigenvalue 1 with non-zero first coordinate. We then rescale it so that the first coordinate is $L_0 = \det(I_{r-1} + H_{r-1})$, and then verify that the sum of the resulting coordinates is the predicted product formula for $\det(I_r + H_r)$. There is no easy way to find such an eigenvector. Fortunately, eigenvectors are easy to verify if you manage to guess correctly.

In the general case where the determinant to be evaluated has polynomial entries, Mills, Robbins, and Rumsey constructed a family of likely candidates for the eigenvector. Though none of their candidates was correct, they were able to find a linear combination of three of their vectors that was. We shall see how they did this in Section 5.3. They confirmed the Macdonald and Andrews conjectures (Conjectures 6 and 7) as well as their own (Conjecture 9) by normalizing this sum of vectors so that its first coordinate was L_0. They then added the coordinates and verified that this yielded the expected polynomial.

Exercises

5.1.1 Prove that the sum of the vectors $(0,\ldots,0,1)$ and

$$\left(\binom{r+1-i}{1-i},\binom{r+2-i}{2-i},\ldots,\binom{2r-i}{r-i}\right),\quad \text{for } 1\le i\le r,$$

is equal to

$$\left(\binom{r+1}{0},\binom{r+2}{1},\ldots,\binom{2r-1}{r-2},1+\binom{2r}{r-1}\right).$$

This is equivalent to proving that, for $1\le k\le r$,

$$\sum_{i=1}^{r}\binom{r+k-i}{k-i} = \binom{r+k}{k-1}.$$

5.1.2 Let M^C be the cofactor matrix of M, and let M^C_{ji} be the entry in row j, column i. Using the definition of the determinant as

$$\det(M) = \sum_{\sigma\in\mathcal{S}_n}(-1)^{\mathcal{I}(\sigma)}\prod_{i=1}^{n}M_{i,\sigma(i)},$$

prove that for any k, $1\le k\le n$,

$$\det(M) = \sum_{i=1}^{n}M_{ki}M^C_{ik}.$$

5.1.3 Prove that if $k\ne j$, then

$$\sum_{i=1}^{n}M_{ki}M^C_{ij} = 0.$$

5.1.4 Prove that the matrix product $M\,M^C$ is the matrix with $\det(M)$ on the diagonal and 0 everywhere else. Use this result to prove that $\det(M^C) = \det(M)^{n-1}$.

5.1.5 Find the inverse matrix for R_5.

5.1.6 Calculate the product $R_5H_5^*R_5^{-1}$ and verify that

$$-R_5H_5^*R_5^{-1} = \left((-1)^j\binom{10-i}{5-i-j}\right)_{i,j=1}^{5}.$$

5.1.7 Verify that $(429, 1287, 2002, 2002, 1287, 429)$ is an eigenvector with eigenvalue 1 for the matrix

$$K_5 = \left((-1)^j\binom{10-i}{5-i-j}\right)_{i,j=0}^{5}.$$

5.1.8 Show that the entry in row i, column j of R is $\binom{r+j-i}{j-i}$. Show that if

$$\sum_{j=1}^{r} \binom{r+j-h}{j-h} \binom{k-j-r-2}{k-j} = \chi(h=k), \tag{5.5}$$

where $\chi(S) = 1$ if S is true and 0 if it is false, then

$$R^{-1} = \left(\binom{j-i-r-2}{j-i} \right)_{i,j=1}^{r}.$$

Note that it is necessary to define

$$\binom{a}{j} = \frac{a(a-1)\cdots(a-j+1)}{j!}$$

when a is negative. Verify that the inverse matrix for R_5 is equal to $\left(\binom{j-i-7}{j-i}\right)$.

5.2 Identities for hypergeometric series

Before we proceed with the proof of Conjecture 6, we need some results from the theory of hypergeometric and basic hypergeometric series. We have seen these series in Chapter 3 where they appeared in the q-binomial theorem, Theorem 3.1, and again in equation (3.28) on page 110. We shall build on these results to obtain transformation and summation formulæ that we need in the next section as well as in the proof of the alternating sign matrix conjecture.

The actual results that are required for the proofs in this book are not particularly difficult or deep. I could have treated them on an ad hoc basis. But hypergeometric series are fundamental to the work now being done in algebraic combinatorics. I have decided to lay a foundation for their study that will facilitate any further explorations you may undertake. Some of the deeper results are treated in the exercises at the end of this section.

We begin by tying up a loose end left from the very start of our story: establishing the equivalence of Conjectures 1 and 2. In Section 1.1, we sought a formula for $A_{n,k}$, the number of $n \times n$ alternating sign matrices with a 1 at the top of the kth column, and we made two conjectures that we asserted to be equivalent. It is easy to see how Conjecture 2,

$$A_{n,k} = \binom{n+k-2}{k-1} \frac{(2n-k-1)!}{(n-k)!} \prod_{j=0}^{n-2} \frac{(3j+1)!}{(n+j)!}, \tag{5.6}$$

implies Conjecture 1,

$$\frac{A_{n,k}}{A_{n,k+1}} = \frac{k(2n-k-1)}{(n-k)(n+k-1)}. \tag{5.7}$$

Using the ratio of consecutive terms given in equation (5.7) to derive the formula of equation (5.6) takes a little more work.

If we know $A_{n,1}$, then we can iterate our ratio to find $A_{n,k}$,

$$\begin{aligned} A_{n,k} &= \frac{(n-1)(n)}{1(2n-2)} \cdot \frac{(n-2)(n+1)}{2(2n-3)} \cdots \frac{(n-k+1)(n+k-2)}{(k-1)(2n-k)} A_{n,1} \\ &= \frac{(n+k-2)!\,(2n-k-1)!}{(n-k)!\,(k-1)!\,(2n-2)!} A_{n,1}. \end{aligned}$$

We can now add up the values of the $A_{n,k}$ to find $A_n = A_{n+1,1}$:

$$A_{n+1,1} = A_n = A_{n,1} \sum_{k=1}^{n} \frac{(n+k-2)!\,(2n-k-1)!}{(n-k)!\,(k-1)!\,(2n-2)!}.$$

Can we express this sum more succinctly?

We can take a bare-hands approach, first rewriting our sum as

$$\begin{aligned} &\sum_{k=1}^{n} \frac{(n+k-2)!\,(2n-k-1)!}{(n-k)!\,(k-1)!\,(2n-2)!} \\ &\qquad = \frac{(n-1)!\,(n-1)!}{(2n-2)!} \sum_{k=1}^{n} \binom{n+k-2}{n-1} \binom{2n-k-1}{n-1}, \end{aligned}$$

and then recognizing the summation as the number of ways of placing $2n-1$ identical objects into distinct positions labeled 1 through $3n-2$. The nth object goes into position $n+k-1$ where $1 \le k \le n$. There are $\binom{n+k-2}{n-1}$ ways of placing the first $n-1$ objects and $\binom{2n-k-1}{n-1}$ ways of placing the last $n-1$ objects. We have proven that

$$\begin{aligned} A_n &= \frac{(n-1)!\,(n-1)!}{(2n-2)!} \binom{3n-2}{2n-1} A_{n-1} \\ &= \frac{(n-1)!\,(3n-2)!}{(2n-2)!\,(2n-1)!} A_{n-1}. \end{aligned}$$

Since $A_1 = 1$, we see that Conjecture 1 implies that

$$A_n = \prod_{k=2}^{n} \frac{(k-1)!\,(3k-2)!}{(2k-2)!\,(2k-1)!}.$$

I leave it as an exercise for you to check that this product equals

$\prod_{j=0}^{n-1}(3j+1)!/(n+j)!$. It follows that

$$\begin{aligned} A_{n,k} &= \frac{(n+k-2)!\,(2n-k-1)!}{(n-k)!\,(k-1)!\,(2n-2)!}\prod_{j=0}^{n-2}\frac{(3j+1)!}{(n+j-1)!} \\ &= \binom{n+k-2}{k-1}\frac{(2n-k-1)!}{(n-k)!}\prod_{j=0}^{n-2}\frac{(3j+1)!}{(n+j)!}. \end{aligned}$$

Generalizing the summation formula

The sum of products of binomial coefficients that we found is a special case of a more general result:

$$\binom{a+b+m+1}{m} = \sum_{k=0}^{m}\binom{a+k}{k}\binom{b+m-k}{m-k}. \tag{5.8}$$

This can be proven using the same counting argument as above (see exercise 5.2.4), or as a consequence of the binomial theorem.

To prove equation (5.8), we first observe that the binomial theorem states that

$$(1-x)^{-n-1} = \sum_{j=0}^{\infty}\binom{n+j}{j}x^j.$$

This implies that

$$\begin{aligned} \sum_{m=0}^{\infty}\binom{a+b+1+m}{m}x^m &= (1-x)^{-a-b-2} \\ &= (1-x)^{-a-1}(1-x)^{-b-1} \\ &= \sum_{k=0}^{\infty}\binom{a+k}{k}x^k \;\sum_{j=0}^{\infty}\binom{b+j}{j}x^j. \end{aligned}$$

We now compare the coefficient of x^m on each side on this equality. On the left, it is the binomial coefficient $\binom{a+b+1+m}{m}$. On the right, k can take on any value less than or equal to m, and then j must equal $m-k$.

The counting argument proof of equation (5.8) is appealing for its simplicity, but the argument just presented is more powerful for it reveals that a and b do not have to be positive integers. In fact, they could be any complex numbers. This leads to problems with our notation. If a and b are negative integers, say $a=-c-1$ and $b=-d-1$, then it is more natural to write the binomial coefficient $\binom{-c-1+k}{k}$ as $(-1)^k\binom{c}{k}$

(see exercise 5.2.5). We see that in this case, k must be less than or equal to the minimum of m, c, and d. Equation (5.8) could appear in the guise

$$\binom{c+d}{m} = \sum_{k\geq 0} \binom{c}{k}\binom{d}{m-k}. \tag{5.9}$$

But there is a deeper problem if we stick to sums of products of binomial coefficients. It suggests that m has to be an integer. In fact, equation (5.8) can be stated in a form in which it is clear that a, b, and m can all be arbitrary complex numbers, subject only to convergence conditions. With the right notation, many summations of products of binomial coefficients reveal themselves to be special cases of this single identity. As we shall see, another special case is

$$\sum_{k=0}^{n} \binom{n+k}{2k}\binom{2k}{k}\frac{(-1)^k}{k+1} = 0, \quad n \geq 1.$$

Standardized notation

The first thing that we do when standardizing the representation is to ensure that the first summand is 1. We do this by factoring out of the summation whatever that first summand might be. Taking equation (5.9) as an example, we divide both sides by $\binom{d}{m}$ so that the identity becomes

$$\sum_{k=0}^{m} \frac{c!\,(d-m)!\,m!}{k!\,(c-k)!\,(d-m+k)!\,(m-k)!} = \frac{(c+d)!\,(d-m)!}{(c+d-m)!\,d!}.$$

This can be simplified if we use the notation of **rising factorial**,

$$(a)_k = \prod_{i=0}^{k-1}(a+i) = a(a+1)\cdots(a+k-1),$$

in the summation, and the gamma function,

$$\Gamma(n) = \int_0^\infty e^{-x}x^{n-1}\,dx \quad (\text{which is } (n-1)! \text{ if } n \in \mathbf{N}),$$

on the product side:

$$\sum_{k=0}^{m} \frac{(c-k+1)_k\,(m-k+1)_k}{k!\,(d-m+1)_k} = \frac{\Gamma(c+d+1)\,\Gamma(d-m+1)}{\Gamma(c+d-m+1)\,\Gamma(d+1)}.$$

We can further standardize this by rewriting each rising factorial so that the index of summation appears only in the subscript:

$$\begin{aligned}(a-k+1)_k &= (a-k+1)(a-k+2)\cdots(a)\\ &= (-1)^k(k-a-1)(k-a-2)\cdots(-a)\\ &= (-1)^k(-a)_k.\end{aligned}$$

Our equation becomes

$$\sum_{k=0}^{m}\frac{(-c)_k\,(-m)_k}{k!\,(d-m+1)_k} = \frac{\Gamma(c+d+1)\,\Gamma(d-m+1)}{\Gamma(c+d-m+1)\,\Gamma(d+1)}.$$

We can run this summation from 0 to ∞ because $(-m)_k$ is zero whenever m is a nonnegative integer and k is larger m. This suggests that this identity might hold for any m. It does, almost. For any complex numbers α, β, and γ for which the real part of $\gamma-\alpha-\beta$ is positive and γ is neither 0 nor a negative integer, we have that

$$\sum_{k=0}^{\infty}\frac{(\alpha)_k\,(\beta)_k}{k!\,(\gamma)_k} = \frac{\Gamma(\gamma-\alpha-\beta)\,\Gamma(\gamma)}{\Gamma(\gamma-\alpha)\,\Gamma(\gamma-\beta)}. \tag{5.10}$$

This result is known as the **Chu–Vandermonde identity**. Chu Shih-Chieh, a fourteenth-century Chinese mathematician, discovered it for integer values of the parameters when at least one of the numerator parameters is negative.† Alexandre Théophile Vandermonde rediscovered this case in the eighteenth century (1772). The general form given here is a special case of a result first stated by Johann Friedrich Pfaff in 1797. Note that if α is a negative integer, $\alpha=-m$, then the ratio of gamma functions can be written as a ratio of rising factorials:

$$\sum_{k=0}^{m}\frac{(-m)_k\,(\beta)_k}{k!\,(\gamma)_k} = \frac{(\gamma-\beta)_m}{(\gamma)_m}. \tag{5.11}$$

Examples

We begin with the example that I said we already know how to sum. It is 1 when $k=0$, so our first step is to rewrite it using rising factorials:

$$\sum_{k=0}^{n}\binom{n+k}{2k}\binom{2k}{k}\frac{(-1)^k}{k+1} = \sum_{k=0}^{n}\frac{(n-k+1)_{2k}\,(-1)^k}{k!\,(k+1)!}.$$

† It appears in *Ssu Yuan Yü Chien (Precious Mirror of the Four Elements)* and is referenced in Askey (1975, p. 59).

The rising factorial in the numerator breaks into two pieces:

$$(n-k+1)_{2k} = (n-k+1)_k\,(n+1)_k = (-1)^k\,(-n)_k\,(n+1)_k.$$

We can also rewrite $(k+1)!$ as $(2)_k$. This summation is now in the desired form, and we can apply equation (5.11):

$$\sum_{k=0}^{n}\binom{n+k}{2k}\binom{2k}{k}\frac{(-1)^k}{k+1} = \sum_{k=0}^{n}\frac{(-n)_k\,(n+1)_k}{k!\,(2)_k} = \frac{(1-n)_n}{(2)_n}.$$

We notice that

$$(1-n)_n \;=\; (1-n)(2-n)\cdots(-1)(0) \;=\; 0,$$

whenever $n \geq 1$. We have shown that

$$\sum_{k=0}^{n}\binom{n+k}{2k}\binom{2k}{k}\frac{(-1)^k}{k+1} \;=\; 0, \quad n \geq 1.$$

As a second example, consider

$$\sum_{k=0}^{n} k\binom{m-k}{m-n}.$$

This is 0 when $k = 0$, and we cannot turn that into 1. This means that we want to re-index, $k = j+1$, and sum from $j = 0$ to $j = n-1$. Our first step is to factor out $\binom{m-1}{m-n}$:

$$\begin{aligned}\sum_{k=0}^{n} k\binom{m-k}{m-n} &= \binom{m-1}{m-n}\sum_{j=0}^{n-1}\frac{(m-1-j)!\,(n-1)!}{(m-1)!\,(n-1-j)!}\,(j+1)\\ &= \binom{m-1}{m-n}\sum_{j=0}^{n-1}\frac{(n-j)_j}{(m-j)_j}\,(j+1).\end{aligned}$$

In addition to rewriting $(n-j)_j/(m-j)_j$ as $(1-n)_j/(1-m)_j$, we need to put a $j!$ in the denominator and the numerator. Again, we write $(j+1)!$ as $(2)_j$:

$$\begin{aligned}\sum_{k=0}^{n} k\binom{m-k}{m-n} &= \binom{m-1}{m-n}\sum_{j=0}^{n-1}\frac{(1-n)_j\,(2)_j}{j!\,(1-m)_j}\\ &= \binom{m-1}{m-n}\frac{(-1-m)_{n-1}}{(1-m)_{n-1}}.\end{aligned}$$

I leave it for you to verify that this is equal to $\binom{m+1}{n-1}$.

Hypergeometric notation

The Chu–Vandermonde summation is an example of a series for which the ratio of consecutive summands is a rational function in the index of summation. In this particular case, we have

$$\frac{(\alpha)_{k+1}\,(\beta)_{k+1}}{(k+1)!\,(\gamma)_{k+1}} \bigg/ \frac{(\alpha)_k\,(\beta)_k}{k!\,(\gamma)_k} = \frac{(k+\alpha)(k+\beta)}{(k+1)(k+\gamma)}.$$

Not only are such series common, there are sharp criteria, proven by Gauss in 1812, for whether such a series diverges, converges conditionally, or converges absolutely (Bressoud 1994, sec. 4.3). For the series given in equation (5.10) with real parameters, we have divergence if $\alpha + \beta \geq 1 + \gamma$, conditional convergence if $1 + \gamma > \alpha + \beta \geq \gamma$, and absolute convergence if $\gamma > \alpha + \beta$.

Hypergeometric series are expressed in a standard notation:

$${}_rF_s\left[\begin{matrix}\alpha_1, \alpha_2, \ldots, \alpha_r \\ \gamma_1, \gamma_2, \ldots, \gamma_s\end{matrix}; x\right] = \sum_{k=0}^{\infty} \frac{(\alpha_1)_k\,(\alpha_2)_k \ldots (\alpha_r)_k}{k!\,(\gamma_1)_k\,(\gamma_2)_k \ldots (\gamma_s)_k}\, x^k.$$

The first subscript on the F tells us the number of parameters in the numerator; the second subscript is the number of parameters in the denominator.

For an arbitrary hypergeometric series, the ratio of the $k+1$st summand to the kth is

$$\frac{(\alpha_1 + k)(\alpha_2 + k)\cdots(\alpha_r + k)\,x}{(\gamma_1 + k)\cdots(\gamma_s + k)}.$$

The parameters α_i and γ_j can be complex. It follows that hypergeometric series are precisely those series for which the ratio of consecutive summands is a rational function of the index of summation. Furthermore, given such a series, we can turn it into standard hypergeometric form by dividing out the summand at $k = 0$ and then finding the rational function that expresses the ratio of consecutive summands. The hypergeometric parameters are the negatives of the roots of the polynomials in this rational function. One of the roots in the denominator must be -1 (which arises from $k!$). If it is not present, then we must multiply the numerator and denominator polynomials by $(1 + k)$.

For example, given

$$\sum_{k\geq 0} 2^k \binom{2m+k}{m+k+1} \bigg/ \binom{2m+2k+1}{m+k+1},$$

the ratio of summands is

$$\frac{\binom{2m+k+1}{m+k+2}}{\binom{2m+2k+3}{m+k+2}} \cdot \frac{\binom{2m+2k+1}{m+k+1}}{\binom{2m+k}{m+k+1}}$$

$$= \frac{(2m+k+1)(m+k+2)(m+k+1)}{(m+k+2)(2m+2k+3)(2m+2k+2)} 2$$

$$= \frac{(1+k)(2m+1+k)}{(1+k)((m+3/2)+k)} (1/2).$$

The standard form for our summation is

$$\sum_{k\geq 0} 2^k \binom{2m+k}{m+k+1} \Big/ \binom{2m+2k+1}{m+k+1} = \frac{m}{2m+1}\, {}_2F_1\left[\begin{matrix} 1,\ 2m+1 \\ m+3/2 \end{matrix} ; \frac{1}{2}\right].$$

We look for an identity that will enable us to sum this.† Gauss's identity,

$${}_2F_1\left[\begin{matrix} a,\ b \\ (a+b+1)/2 \end{matrix} ; \frac{1}{2}\right] = \frac{\Gamma(1/2)\,\Gamma((a+b+1)/2)}{\Gamma((a+1)/2)\,\Gamma((b+1)/2)},$$

finishes the problem (see exercise 5.2.13).

Return to the q-binomial theorem

We shall use the q-binomial theorem, Theorem 3.3 on page 78, to prove a q-analog of the Chu–Vandermonde identity, a more general identity that becomes equation (5.10) in the limit as q approaches 1. Introducing this extra parameter simplifies the proof. It also yields identities that we shall need in the next section.

We observe that the summation side of the binomial theorem can be expressed using hypergeometric notation:

$$\sum_{k=0}^{\infty} \frac{\alpha(\alpha-1)\cdots(\alpha-k+1)}{k!} x^k = \sum_{k=0}^{\infty} \frac{(-\alpha)_k}{k!} (-x)^k$$

$$= {}_1F_0\left[\begin{matrix} -\alpha \\ \ \end{matrix} ; -x\right].$$

The general binomial theorem states that

$${}_1F_0\left[\begin{matrix} \alpha \\ \ \end{matrix} ; x\right] = (1-x)^{-\alpha} \qquad (5.12)$$

† Christian Krattenthaler's HYP.m is a *Mathematica* package that, among other things, searches for appropriate identities for hypergeometric series. It can be found at radon.mat.univie.ac.at/People/kratt/.

for any complex number α.

We shall need something analogous to the rising factorial if we are to rewrite the q-binomial theorem in a similar form. We introduce the **rising q-factorial**:

$$(a;q)_n = (1-a)(1-aq)\cdots(1-aq^{n-1}),$$

where n is any positive integer. If the absolute value of q is less than 1, then an infinite product of this form will converge for any complex number a (see exercise 5.2.11):

$$(a;q)_\infty = \prod_{i=0}^{\infty}(1-aq^i).$$

For any complex number β, we define

$$(a;q)_\beta = \frac{(a;q)_\infty}{(aq^\beta;q)_\infty}, \qquad |q|<1.$$

We observe that this is consistent with the definition given above. When n is a positive integer, we have

$$\frac{(a;q)_\infty}{(aq^n;q)_\infty} = (1-a)(1-aq)\cdots(1-aq^{n-1}).$$

If we replace x by $-x/q$ in Theorem 3.3, the q-binomial theorem can be written using rising q-factorial notation as

$$(x;q)_n = \sum_{k=0}^{n} \frac{(1-q^n)(1-q^{n-1})\cdots(1-q^{n-k+1})}{(1-q)\cdots(1-q^k)}\, q^{k(k-1)/2}(-x)^k.$$

We factor $-q^{n-i}$, $0\le i<k$, out of each of the binomials in the numerator so that it becomes

$$(1-q^{-n})(1-q^{-n+1})\cdots(1-q^{-n+k-1})(-1)^k q^{nk-k(k-1)/2}.$$

Theorem 3.3 can be expressed as

$$\frac{(x;q)_\infty}{(xq^n;q)_\infty} = \sum_{k=0}^{n} \frac{(q^{-n};q)_k}{(q;q)_k}\,(xq^n)^k.$$

This suggests the following theorem which has a simple proof.

Theorem 5.1 (The general q-binomial theorem) *For complex values of x, b, and q where $|x|<1$ and $|q|<1$, we have that*

$$\frac{(xb;q)_\infty}{(x;q)_\infty} = \sum_{k=0}^{\infty} \frac{(b;q)_k}{(q;q)_k}\, x^k. \tag{5.13}$$

Proof: We begin with the rational product $(xb;q)_\infty/(x;q)_\infty$. This is an analytic function in x for $|x|<1$, and so it has a power series expansion:

$$f(x) = \frac{(xb;q)_\infty}{(x;q)_\infty} = \sum_{k=0}^{\infty} a_k\, x^k.$$

We want to find a relationship between consecutive coefficients. We observe that

$$f(xq) = \frac{(xbq;q)_\infty}{(xq;q)_\infty} = \frac{(1-x)}{(1-xb)}\, f(x).$$

This implies that

$$(1-xb)f(xq) = (1-x)f(x),$$

which is the same as

$$\begin{aligned}(1-xb)\sum_{k=0}^{\infty} a_k\, x^k q^k &= (1-x)\sum_{k=0}^{\infty} a_k\, x^k, \\ \sum_{k=0}^{\infty} a_k\, x^k q^k - b\sum_{k=0}^{\infty} a_k\, x^{k+1} q^k &= \sum_{k=0}^{\infty} a_k\, x^k - \sum_{k=0}^{\infty} a_k\, x^{k+1}.\end{aligned}$$

We now compare the coefficients of x^n, $n \ge 1$, on each side:

$$a_n q^n - b a_{n-1} q^{n-1} = a_n - a_{n-1}.$$

We solve this equation for a_n in terms of a_{n-1}:

$$\begin{aligned}a_n(q^n-1) &= a_{n-1}(bq^{n-1}-1), \\ a_n &= \frac{(1-bq^{n-1})}{(1-q^n)}\, a_{n-1}.\end{aligned}$$

We can iterate this relationship to write a_n in terms of a_{n-2} and continue until a_n is expressed as a multiple of a_0:

$$\begin{aligned}a_n &= \frac{(1-bq^{n-1})(1-bq^{n-2})}{(1-q^n)(1-q^{n-1})}\, a_{n-2} \\ &\vdots \\ &= \frac{(1-bq^{n-1})(1-bq^{n-2})\cdots(1-b)}{(1-q^n)(1-q^{n-1})\cdots(1-q)}\, a_0 \\ &= \frac{(b;q)_n}{(q;q)_n}\, a_0.\end{aligned}$$

We observe that a_0 is the constant term in our product:

$$a_0 = \frac{(0\cdot b;q)_\infty}{(0;q)_\infty} = 1,$$

and therefore

$$a_n = \frac{(b;q)_n}{(q;q)_n}.$$

Q.E.D.

Basic hypergeometric notation

In exact analogy with the hypergeometric notation for a sum that involves rising factorials, there is an analog that involves rising q-factorials called **basic hypergeometric series**:

$$ {}_2\phi_1 \left[\begin{matrix} a, b \\ c \end{matrix} ; q; x \right] = {}_2\phi_1 \left[\begin{matrix} a, b \\ c \end{matrix} ; x \right] = \sum_{k=0}^{\infty} \frac{(a;q)_k \, (b;q)_k}{(q;q)_k \, (c;q)_k} x^k. $$

This converges for $|q| < 1, |x| < 1, c \neq 1, q^{-1}, q^{-2}, \ldots$. When the **base** is simply q, we usually do not write it. In the proof of the alternating sign matrix conjecture in Chapter 7, we shall need to replace the base q by q^3:

$$ {}_2\phi_1 \left[\begin{matrix} a, b \\ c \end{matrix} ; q^3; x \right] = \sum_{k=0}^{\infty} \frac{(a;q^3)_k \, (b;q^3)_k}{(q^3;q^3)_k \, (c;q^3)_k} x^k. $$

Both our notation and many fundamental results for basic hypergeometric series come from Heinrich Eduard Heine (1846, 1847), a student of Dirichlet who is best known for his work in analysis. Of particular importance are Heine's transformation formulæ.

Theorem 5.2 (Heine's transformation formulæ) *Provided the series in question converge, we have that*

$$ {}_2\phi_1 \left[\begin{matrix} a, b \\ c \end{matrix} ; x \right] = \frac{(b;q)_\infty (ax;q)_\infty}{(c;q)_\infty (x;q)_\infty} \, {}_2\phi_1 \left[\begin{matrix} x, c/b \\ ax \end{matrix} ; b \right], \tag{5.14} $$

$$ = \frac{(c/b;q)_\infty (bx;q)_\infty}{(c;q)_\infty (x;q)_\infty} \, {}_2\phi_1 \left[\begin{matrix} b, abx/c \\ bx \end{matrix} ; c/b \right], \tag{5.15} $$

$$ = \frac{(abx/c;q)_\infty}{(x;q)_\infty} \, {}_2\phi_1 \left[\begin{matrix} c/b, c/a \\ c \end{matrix} ; \frac{abx}{c} \right]. \tag{5.16} $$

Proof: We observe that the second and third transformations follow from the first. We switch the two numerator parameters, and then reapply equation (5.14). The details are left for exercise 5.2.12.

We begin the proof of equation (5.14) with the observation that

$$(b;q)_k = \frac{(b;q)_\infty}{(bq^k;q)_\infty},$$

from which we can rewrite our basic hypergeometric series as

$$\sum_{k=0}^{\infty} \frac{(a;q)_k\,(b;q)_k}{(q;q)_k\,(c;q)_k}\,x^k \;=\; \frac{(b;q)_\infty}{(c;q)_\infty} \sum_{k=0}^{\infty} \frac{(a;q)_k x^k}{(q;q)_k}\,\frac{(cq^k;q)_\infty}{(bq^k;q)_\infty}.$$

We now use equation (5.13) to expand $(cq^k;q)_\infty/(bq^k;q)_\infty$ and then interchange the order of summation (this is allowed because we have absolute convergence):

$$\begin{aligned}
\sum_{k=0}^{\infty} \frac{(a;q)_k\,(b;q)_k}{(q;q)_k\,(c;q)_k}\,x^k &= \frac{(b;q)_\infty}{(c;q)_\infty} \sum_{k=0}^{\infty} \frac{(a;q)_k x^k}{(q;q)_k} \sum_{j=0}^{\infty} \frac{(c/b;q)_j}{(q;q)_j}\,b^j q^{jk} \\
&= \frac{(b;q)_\infty}{(c;q)_\infty} \sum_{j=0}^{\infty} \frac{(c/b;q)_j}{(q;q)_j}\,b^j \sum_{k=0}^{\infty} \frac{(a;q)_k (xq^j)^k}{(q;q)_k}.
\end{aligned}$$

We use equation (5.13) to simplify the sum on k:

$$\sum_{k=0}^{\infty} \frac{(a;q)_k\,(b;q)_k}{(q;q)_k\,(c;q)_k}\,x^k = \frac{(b;q)_\infty}{(c;q)_\infty} \sum_{j=0}^{\infty} \frac{(c/b;q)_j}{(q;q)_j}\,b^j\,\frac{(axq^j;q)_\infty}{(xq^j;q)_\infty}. \tag{5.17}$$

We now rewrite

$$\frac{(axq^j;q)_\infty}{(xq^j;q)_\infty} = \frac{(ax;q)_\infty}{(x;q)_\infty}\,\frac{(x;q)_j}{(ax;q)_j}.$$

Q.E.D.

Theorem 5.2 has several consequences, including the q-analog of the Chu–Vandermonde identity.

Corollary 5.3 *The q-analog of the Chu–Vandermonde identity is*

$${}_2\phi_1\left[\begin{matrix} a, b \\ c \end{matrix}\,;c/ab\right] = \frac{(c/a;q)_\infty\,(c/b;q)_\infty}{(c;q)_\infty\,(c/ab;q)_\infty}, \tag{5.18}$$

provided $|c/ab| < 1$ and $|q| < 1$. We also have the following finite summation formulæ for any positive integer, t:

$$\begin{aligned}
{}_2\phi_1\left[\begin{matrix} q^{-t}, b \\ c \end{matrix}\,;q\right] &= \frac{(bq^{1-t}/c;q)_t}{(q^{1-t}/c;q)_t}, & (5.19)\\
\sum_{k=0}^{t} \frac{(a;q)_k}{(q;q)_k}\,a^{-k} &= a^{-t}\,\frac{(aq;q)_t}{(q;q)_t}. & (5.20)
\end{aligned}$$

In exercise 5.2.10, I ask you to prove that equation (5.18) implies the Chu–Vandermonde identity, equation (5.10).

Proof: To prove the q-analog of Chu–Vandermonde, we set $x = c/ab$ in equation (5.15). The rising q-factorial, $(1;q)_k$, is zero for $k \geq 1$, and therefore

$$ {}_2\phi_1\left[\begin{matrix} b,\ 1 \\ c/a \end{matrix} ; \frac{c}{b}\right] = 1. $$

To prove equation (5.19), we set $a = q^{-t}$ and $x = q$ in equation (5.17). All summands will be zero for k larger than t.

$$ \sum_{k=0}^{t} \frac{(q^{-t};q)_k\,(b;q)_k}{(q;q)_k\,(c;q)_k}\, q^k = \frac{(b;q)_\infty}{(c;q)_\infty} \sum_{j=0}^{\infty} \frac{(c/b;q)_j}{(q;q)_j}\, b^j\, \frac{(q^{j+1-t};q)_\infty}{(q^{j+1};q)_\infty}. $$

The summands on the right are zero unless j is greater than or equal to t. We can replace the index j by $j+t$ and rewrite $(c/b;q)_{j+t}$ as $(c/b;q)_t (cq^t/b;q)_j$:

$$ \begin{aligned} \sum_{k=0}^{t} \frac{(q^{-t};q)_k\,(b;q)_k}{(q;q)_k\,(c;q)_k}\, q^k &= \frac{(b;q)_\infty}{(c;q)_\infty} \sum_{j=0}^{\infty} \frac{(c/b;q)_{j+t}}{(q;q)_\infty}\, b^{j+t}\, (q^{j+1};q)_\infty \\ &= \frac{(b;q)_\infty\,(c/b;q)_t}{(c;q)_\infty}\, b^t \sum_{j=0}^{\infty} \frac{(cq^t/b;q)_j}{(q;q)_j}\, b^j. \end{aligned} $$

We use equation (5.13) to write the sum on j as a product:

$$ \begin{aligned} \sum_{k=0}^{t} \frac{(q^{-t};q)_k\,(b;q)_k}{(q;q)_k\,(c;q)_k}\, q^k &= \frac{(b;q)_\infty\,(c/b;q)_t}{(c;q)_\infty}\, b^t\, \frac{(cq^t;q)_\infty}{(b;q)_\infty} \\ &= b^t\, \frac{(c/b;q)_t}{(c;q)_t}. \end{aligned} $$

To put this in the desired form, we use the fact (see exercise 5.2.14) that

$$ (A;q)_t = (-A)^t\, q^{t(t-1)/2}\, (A^{-1}q^{1-t};q)_t, \tag{5.21} $$

to turn $b^t (c/b;q)_t/(c;q)_t$ into $(bq^{1-t}/c;q)_t/(q^{1-t}/c;q)_t$.

If we set $b = q^{-t}$ and $c = yq^{-t}$ in equation (5.18), then it becomes

$$ \begin{aligned} \sum_{i=0}^{t} \frac{(a;q)_i\,(q^{-t};q)_i}{(q;q)_i\,(yq^{-t};q)_i} \left(\frac{y}{a}\right)^i &= \frac{(ya^{-1}q^{-t};q)_\infty\,(y;q)_\infty}{(yq^{-t};q)_\infty\,(ya^{-1};q)_\infty} \\ &= \frac{(ya^{-1}q^{-t};q)_t}{(yq^{-t};q)_t}. \end{aligned} $$

Equation (5.21) allows us to rewrite this equality as

$$\sum_{i=0}^{t} \frac{(a;q)_i\,(q^{-t};q)_i}{(q;q)_i\,(yq^{-t};q)_i} \left(\frac{y}{a}\right)^i = a^{-t}\,\frac{(ay^{-1}q;q)_t}{(y^{-1}q;q)_t}. \tag{5.22}$$

Equation (5.20) now follows when we set $y = 1$.

Q.E.D.

Exercises

5.2.1 Verify that if

$$A_{n,k} = \binom{n+k-2}{k-1} \frac{(2n-k-1)!}{(n-k)!} \prod_{j=0}^{n-2} \frac{(3j+1)!}{(n+j)!},$$

then

$$\frac{A_{n,k}}{A_{n,k+1}} = \frac{k(2n-k-1)}{(n-k)(n+k-1)}.$$

5.2.2 Prove that

$$\prod_{k=2}^{n} \frac{(k-1)!\,(3k-2)!}{(2k-2)!\,(2k-1)!} = \prod_{j=0}^{n-1} \frac{(3j+1)!}{(n+j)!}.$$

5.2.3 Prove that

$$\frac{(n+k-2)!\,(2n-k-1)!}{(n-k)!\,(k-1)!\,(2n-2)!} \prod_{j=0}^{n-2} \frac{(3j+1)!}{(n+j-1)!}$$
$$= \binom{n+k-2}{k-1} \frac{(2n-k-1)!}{(n-k)!} \prod_{j=0}^{n-2} \frac{(3j+1)!}{(n+j)!}.$$

5.2.4 Prove equation (5.8) by showing that each side counts the number of ways of choosing $a+b+1$ positions from a choice of $a+b+m+1$ positions.

5.2.5 Prove that

$$\binom{-c-1-k}{k} = (-1)^k \binom{c}{k}.$$

5.2.6 Find as many different proofs as possible that

$$\binom{c+d}{m} = \sum_{k=0}^{m} \binom{c}{k} \binom{d}{m-k}.$$

5.2.7 Prove that

$$\binom{m-1}{m-n}\frac{(-1-m)_{n-1}}{(1-m)_{n-1}} = \binom{m+1}{n-1}.$$

5.2.8 Find a closed form for the value of

$$\sum_{k\geq 0}\binom{m}{k}\binom{n}{k}k.$$

5.2.9 Find a closed form for

$$\sum_{k\geq 0}\binom{n}{k}\binom{2k}{k}(-1/4)^k.$$

5.2.10 Prove that equation (5.18) implies the Chu–Vandermonde identity, equation (5.10). The first step is to verify that

$$\lim_{q\to 1^-}\frac{(q^\alpha;q)_k\,(q^\beta;q)_k}{(q;q)_k\,(q^\gamma;q)_k}\,q^{k(\gamma-\alpha-\beta)} = \frac{(\alpha)_k(\beta)_k}{k!\,(\gamma)_k}.$$

Now use Euler's identity (Whittaker and Watson 1927, p. 237),

$$\Gamma(z) = \lim_{n\to\infty}\frac{(n-1)!}{(z)_n}\,n^z,$$

to prove that

$$\begin{aligned}
&\lim_{q\to 1^-}\frac{(q^{\gamma-\alpha};q)_\infty\,(q^{\gamma-\beta};q)_\infty}{(q^\gamma;q)_\infty\,(q^{\gamma-\alpha-\beta};q)_\infty}\\
&\quad= \lim_{n\to\infty}\lim_{q\to 1^-}\frac{(q^{\gamma-\alpha};q)_n\,(q^{\gamma-\beta};q)_n}{(q^\gamma;q)_n\,(q^{\gamma-\alpha-\beta};q)_n}\\
&\quad= \lim_{n\to\infty}\frac{(\gamma-\alpha)_n\,(\gamma-\beta)_n}{(\gamma-\alpha-\beta)_n\,(\gamma)_n}\\
&\quad= \frac{\Gamma(\gamma-\alpha-\beta)\,\Gamma(\gamma)}{\Gamma(\gamma-\alpha)\,\Gamma(\gamma-\beta)}.
\end{aligned}$$

5.2.11 Prove that $\prod_{i=0}^{\infty}(1-aq^i)$, $|q|<1$, converges for any complex number a. An infinite product, $\prod \alpha_i$, converges if and only if $\sum \ln \alpha_i$ converges. In general, one has to worry about the branch of the natural logarithm. Show that for i sufficiently large, the real part of $1-aq^i$ is strictly positive, so we can take the imaginary part of the natural logarithm between $-i\pi/2$ and $i\pi/2$.

5.2.12 Use equation (5.14) to prove equations (5.15) and (5.16). Switch numerator parameters and repeat the transformation given by equation (5.14). What happens if you continue this procedure?

5.2.13 Finish the proof that

$$\sum_{k\geq 0} 2^k \binom{2m+k}{m+k+1} \Big/ \binom{2m+2k+1}{m+k+1} = \frac{\pi}{4} \cdot \frac{3 \cdot 5 \cdots (2m-1)}{2 \cdot 4 \cdots (2m-2)}.$$

5.2.14 Prove equation (5.21). **Hint:** Write each binomial, $1 - Aq^i$, as $-Aq^i(1 - A^{-1}q^{-i})$. Use this to prove that $b^t(c/b;q)_t/(c;q)_t$ equals $(bq^{1-t}/c;q)_t/(q^{1-t}/c;q)_t$.

5.2.15 Prove that

$$\sum_{k=0}^{t} \binom{m+k}{k} = \binom{m+t+1}{t}. \tag{5.23}$$

is a corollary of equation (5.20) by setting $a = q^{m+1}$ and then taking the limit $q \to 1$.

5.2.16 Use equation (5.20) to prove equation (3.28) on page 110:

$$\sum_{k=1}^{r} q^{k(r-1)} \begin{bmatrix} r-2+j-k \\ j-k \end{bmatrix} = q^{r-1} \begin{bmatrix} r-2+j \\ j-1 \end{bmatrix}, \quad 1 \leq j \leq r.$$

Hint: The summation actually goes from $k = 1$ to $k = j$ since all summands are zero for $k > j$. Set $k = j - i$ where i, the new index of summation, ranges from 0 to $j - 1$.

5.2.17 Prove the following identity that was discovered by F. H. Jackson in 1910:

$${}_3\phi_2 \left[\begin{matrix} c/a, \ c/b, \ q^{-n} \\ c, \ cq^{1-n}/ab \end{matrix} ; q \right] = \frac{(a;q)_n \, (b:q)_n}{(c;q)_n \, (ab/c;q)_n}. \tag{5.24}$$

This is an example of a formula for a **balanced** basic hypergeometric series: a series for which the product of the numerator parameters times the argument, $(c/a)(c/b)q^{-n} \cdot q$, is equal to the product of the denominator parameters, $c(cq^{1-n}/ab)$. Note that the q-analog of Chu–Vandermonde, equation (5.18), is also an identity for a balanced series.

Hint: Start with equation (5.16). Use equation (5.13) to expand $(abx/c;q)_\infty/(x;q)_\infty$ as a power series in x. Compare the coefficient of x^n on each side.

5.2.18 This exercise will lead you through a proof of the **Rogers–Ramanujan identities**, equations (5.26) and (5.27). Show that

$$\sum_{k=0}^{n-m} \frac{q^{k^2+2mk}(q;q)_{n-m}}{(q;q)_{k+2m}\,(q;q)_k\,(q;q)_{n-m-k}} = \frac{1}{(q;q)_{2m}} \lim_{b\to\infty} {}_2\phi_1\left[\begin{matrix} q^{m-n},\ b \\ q^{2m+1} \end{matrix} ; \frac{q^{m+n+1}}{b}\right].$$

Use equation (5.18) to prove that

$$\sum_{k=0}^{n-m} \frac{q^{k^2+2mk}\,(q;q)_{n-m}}{(q;q)_{k+2m}\,(q;q)_k\,(q;q)_{n-m-k}} = \frac{1}{(q;q)_{m+n}}.$$

Now use this result to prove that

$$\begin{aligned}
\sum_{m=-n}^{n} \frac{x^m\,q^{am^2}}{(q;q)_{n-m}\,(q;q)_{n+m}}
&= \sum_{m=-n}^{n}\sum_{k=0}^{n-m} \frac{q^{(m+k)^2}}{(q;q)_{n-m-k}} \frac{x^m q^{(a-1)m^2}}{(q;q)_k\,(q;q)_{2m+k}} \\
&= \sum_{s=0}^{n} \frac{q^{s^2}}{(q;q)_{n-s}} \sum_{m=-s}^{s} \frac{x^m q^{(a-1)m^2}}{(q;q)_{s-m}\,(q;q)_{s+m}}. \qquad (5.25)
\end{aligned}$$

Iterate equation (5.25) and then use equation (3.6) on page 80 to prove that

$$\sum_{m=-n}^{n} \frac{(xq^{1/2})^m\,q^{5m^2/2}}{(q;q)_{n-m}\,(q;q)_{n+m}} = \sum_{0\le t\le s\le n} \frac{q^{s^2+t^2}\,(-xq;q)_t\,(-x^{-1};q)_t}{(q;q)_{n-s}\,(q;q)_{s-t}\,(q;q)_{2t}}.$$

Take the limit as n approaches ∞ and use the Jacobi triple product identity, equation (2.11), to prove that, for $x=-1$:

$$\prod_{j=1}^{\infty} \frac{1}{(1-q^{5j-4})(1-q^{5j-1})} = \sum_{s=0}^{\infty} \frac{q^{s^2}}{(q;q)_s}, \qquad (5.26)$$

and, for $x=-q$:

$$\prod_{j=1}^{\infty} \frac{1}{(1-q^{5j-3})(1-q^{5j-2})} = \sum_{s=0}^{\infty} \frac{q^{s^2+s}}{(q;q)_s}. \qquad (5.27)$$

The second identity is slightly tricky. The index t can be 0 or

1, from which it follows that the right-hand side is equal to

$$\sum_{s=0}^{\infty} \frac{q^{s^2}}{(q;q)_s} - \sum_{s=1}^{\infty} \frac{q^{s^2}}{(q;q)_{s-1}}.$$

5.3 Proof of the Macdonald conjecture

We are now prepared to fill in the remaining details of the Mills, Robbins, and Rumsey proof of Conjecture 6, that the generating function for cyclically symmetric plane partitions that fit inside $\mathcal{B}(r,r,r)$ is given by

$$\prod_{\eta \in \mathcal{B}(r,r,r)/\mathcal{C}_3} \frac{1 - q^{|\eta|(1+\mathrm{ht}(\eta))}}{1 - q^{|\eta|\,\mathrm{ht}(\eta)}}.$$

Actually, what they did was to prove a far more general result that encompasses Conjectures 6, 7, and 9. This includes the result that the generating function for descending plane partitions that fit inside the same $r \times r \times r$ box is given by

$$\prod_{1 \le i \le j \le r} \frac{1 - q^{r+i+j-1}}{1 - q^{2i+j-1}}.$$

More than this, we shall follow Mills, Robbins, and Rumsey in deriving the generating functions for cyclically symmetric plane partitions and descending plane partitions when we keep track of the number of parts of maximum size in the bottom level.

As we saw in Chapter 3, Theorems 3.10 and 3.11, our generating functions can be represented by determinants. Conjectures 6 and 7 are equivalent to the following determinant evaluations:

$$\det\left(\delta_{ij} + q^{3i-2}\begin{bmatrix} i+j-2 \\ i-1 \end{bmatrix}_{q^3}\right)_{i,j=1}^{r} = \prod_{\eta \in \mathcal{B}(r,r,r)/\mathcal{C}_3} \frac{1 - q^{|\eta|(1+\mathrm{ht}(\eta))}}{1 - q^{|\eta|\,\mathrm{ht}(\eta)}}, \tag{5.28}$$

$$\det\left(\delta_{ij} + q^{i+1}\begin{bmatrix} i+j \\ i+1 \end{bmatrix}\right)_{i,j=1}^{r} = \prod_{1 \le i \le j \le r+1} \frac{1 - q^{r+i+j}}{1 - q^{2i+j-1}}. \tag{5.29}$$

We shall evaluate the determinant of $I_r + T_r(\lambda, d)$ where

$$T_r(\lambda, d) = \left(\lambda q^{i+1-d}\begin{bmatrix} i+j-d \\ j-1 \end{bmatrix}\right)_{i,j=1}^{r}.$$

There is a relationship between λ and d that will be introduced when it becomes necessary. Macdonald's conjecture (Conjecture 6) will follow from this determinant evaluation with $\lambda = q^{1/3}$, $d = 2$, and q replaced by q^3. Andrews's conjecture (Conjecture 7) will be the special case $\lambda = 1$ and $d = 0$.

As the product form of the determinants suggests, it is reasonable to seek a compact formula for

$$\frac{\det(I_r + T_r)}{\det(I_{r-1} + T_{r-1})}$$

and then complete the proof by induction on r. In the exercises, the reader is asked to verify that equations (5.28) and (5.29) are correct for $r = 1$ and $r = 2$ and that, given these initial conditions, they are equivalent to

$$\frac{\det\left(I_r + T_r(q^{1/3}, 2)\right)}{\det\left(I_{r-1} + T_{r-1}(q^{1/3}, 2)\right)} = \frac{1 - q^{r-1/3}}{1 - q^{r-2/3}} \prod_{i=1}^{r-1} \frac{1 - q^{2r-1+i}}{1 - q^{r-1+i}}, \tag{5.30}$$

$$\frac{\det\left(I_r + T_r(1, 0)\right)}{\det\left(I_{r-1} + T_{r-1}(1, 0)\right)} = \prod_{i=1}^{r} \frac{1 - q^{2r+i+1}}{1 - q^{r+i}}. \tag{5.31}$$

We shall assume that $I_{r-1} + T_{r-1}(\lambda, d)$ is known to be non-singular and then find a general formula for $\det(I_r + T_r(\lambda, d)) / \det(I_{r-1} + T_{r-1}(\lambda, d))$ that includes equations (5.30) and (5.31) as special cases. This will be done in three stages which I sketch below. We shall then fill in the details.

1. We shall find $r + 1$ r-dimensional vectors: $R_0, R_1, \ldots, R_r$ such that $R_0 + R_1 + \cdots + R_r$ is the bottom row of $I_r + T_r$. We let U_k, $0 \le k \le r$ denote $I_r + T_r$ with the last row replaced by R_k, and define $L_k = \det(U_k)$. We specify that $R_0 = (0, \ldots, 0, 1)$, so that

$$\det(I_{r-1} + T_{r-1}) = L_0.$$

   I ask you to prove (exercise 5.3.6) that

$$\det(I_r + T_r) = L_0 + L_1 + \cdots + L_r.$$

   It follows that

$$\frac{\det(I_r + T_r)}{\det(I_{r-1} + T_{r-1})} = 1 + \sum_{k=1}^{r} L_k / L_0. \tag{5.32}$$

   As it happens, L_k is the generating function when we specify that there must be k parts of maximum size. This fact – which was instrumental in discovering the proofs of Conjectures 6 and 7 – is irrelevant to their validity.

2. We let R be the $r \times r$ matrix whose rows are $R_1, R_2, \ldots, R_r$. As we shall see, R is upper triangular with non-zero entries on the main diagonal, so it is non-singular. We shall explicitly compute R^{-1}.

Let C be the column vector of cofactors of the last row of $I_r + T_r$. By the definition of a cofactor, we have that

$$RC = \begin{pmatrix} L_1 \\ L_2 \\ \vdots \\ L_r \end{pmatrix},$$

and therefore

$$RU_0R^{-1} \begin{pmatrix} L_1 \\ L_2 \\ \vdots \\ L_r \end{pmatrix} = RU_0C.$$

The dot product of the ith row of U_0 with C is the determinant of the matrix obtained by replacing the last of U_0 by the ith row of U_0. It follows that

$$\begin{aligned} RU_0R^{-1} \begin{pmatrix} L_1 \\ L_2 \\ \vdots \\ L_r \end{pmatrix} &= R \begin{pmatrix} 0 \\ 0 \\ \vdots \\ 0 \\ L_0 \end{pmatrix} \\ &= L_0 \begin{pmatrix} R_{1,r} \\ R_{2,r} \\ \vdots \\ R_{r,r} \end{pmatrix}, \end{aligned}$$

where $R_{k,r}$ is the rth coordinate of R_k. Since $\det(U_0) = L_0$ which is not zero by our induction hypothesis, this system of equations uniquely determines L_1 through L_r.

Let T_r^* be the matrix T_r with all entries in the bottom row replaced by zeros so that $U_0 = I_r + T_r^*$. Equation (5.33) can be rewritten as

$$\begin{pmatrix} L_1 \\ L_2 \\ \vdots \\ L_r \end{pmatrix} = L_0 \begin{pmatrix} R_{1,r} \\ R_{2,r} \\ \vdots \\ R_{r,r} \end{pmatrix} - RT_r^* R^{-1} \begin{pmatrix} L_1 \\ L_2 \\ \vdots \\ L_r \end{pmatrix}$$

$$= M^* \begin{pmatrix} L_0 \\ L_1 \\ L_2 \\ \vdots \\ L_r, \end{pmatrix} \qquad (5.33)$$

where M^* is the $r \times (r+1)$ matrix whose first column is

$$(R_{1,r}, R_{2,r}, \ldots, R_{r,r})$$

and whose remaining r columns are the matrix $-RT_r^* R^{-1}$. We shall compute the entries of the matrix M^* and observe that there is a natural way of defining an $(r+1) \times (r+1)$ matrix M whose bottom r rows agree with M^*.

3. Equation (5.33) uniquely determines L_1 through L_r. If M has an eigenvalue equal to 1 and there is a corresponding eigenvector with first coordinate 1, it must be $(1, L_1/L_0, \ldots, L_r/L_0)$. We shall find such an eigenvector as well as a compact formula for the sum of its coordinates.

Decomposing the determinant of $I_r + T_r$

In Section 3.4, we found determinants that give the generating functions for cyclically symmetric plane partitions, Theorem 3.10, and for descending plane partitions, Theorem 3.11, when we specify the number of parts of maximal size. These formulæ suggest that we want to decompose the determinant of $I_r + T_j$ by replacing the last row by $R_0 = (0, \ldots, 0, 1)$ or by R_k defined as

$$\lambda q^{k(r+1-d)} \left(\begin{bmatrix} r-d+1-k \\ 1-k \end{bmatrix}, \begin{bmatrix} r-d+2-k \\ 2-k \end{bmatrix}, \ldots, \begin{bmatrix} r-d+r-k \\ r-k \end{bmatrix} \right)$$

for $1 \le k \le r$. We need to prove that the sum of these rows is equal to the last row of $I_r + T_r$ which means that we must prove that for each coordinate j, $1 \le j \le r$, we have

$$\lambda \sum_{k=1}^{r} q^{k(r+1-d)} \begin{bmatrix} r-d+j-k \\ j-k \end{bmatrix} = \lambda q^{r+1-d} \begin{bmatrix} r+j-d \\ j-1 \end{bmatrix}.$$

Since j is less than or equal to r and $k > j$ implies that the summand is zero, the summation only goes as far as $k = j$. We replace k by $j-k$. We want to prove that

$$\lambda \sum_{k=0}^{j-1} q^{(j-k)(r+1-d)} \begin{bmatrix} r-d+k \\ k \end{bmatrix} = \lambda q^{r+1-d} \begin{bmatrix} r+j-d \\ j-1 \end{bmatrix}.$$

To put this into the form of a hypergeometric series, we divide each side by the first term in the series: $\lambda q^{j(r+1-d)}$. The equality that we need to prove is

$$\sum_{k=0}^{j-1} \frac{(q^{r-d+1};q)_k}{(q;q)_k} q^{k(d-r-1)} = q^{(j-1)(d-r-1)} \frac{(q^{r-d+2};q)_{j-1}}{(q;q)_{j-1}}.$$

This is equation (5.20) with $t = j-1$ and $a = q^{r-d+1}$.

Calculation of R^{-1}

The matrix

$$R = \left(\lambda q^{i(r+1-d)} \begin{bmatrix} r-d+j-i \\ j-i \end{bmatrix} \right)_{i,j=1}^{r}$$

is an upper triangular matrix with non-zero entries on the diagonal provided $\lambda q \neq 0$. It has an inverse of the same form which can be calculated explicitly for small values of r. From here it is not too difficult to guess that the general inverse matrix is

$$R^{-1} = \left(\lambda^{-1} q^{i(d-r-1)} \begin{bmatrix} d-r-2+j-i \\ j-i \end{bmatrix} \right)_{i,j=1}^{r}.$$

To verify that this is the correct inverse, we shall calculate the entry, $a_{h,k}$, in row h, column k of the product of these two matrices:

$$\sum_{j=1}^{r} \lambda q^{h(r+1-d)} \begin{bmatrix} r-d+j-h \\ j-h \end{bmatrix} \lambda^{-1} q^{j(d-r-1)} \begin{bmatrix} d-r-2+k-j \\ k-j \end{bmatrix}.$$

We observe that these summands are zero unless $h \le j \le k$, and in particular $a_{h,k}$ is zero if k is strictly less than h. We replace the index of summation j by $j+h$ and simplify our summands:

$$a_{h,k} = \sum_{j=0}^{k-h} q^{j(d-r-1)} \begin{bmatrix} r-d+j \\ j \end{bmatrix} \begin{bmatrix} d-r-2+k-h-j \\ k-h-j \end{bmatrix}, \qquad h \le k.$$

We need to show that this sum is 0 if $k > h$ and 1 if $k = h$. Our first step is to write it as a basic hypergeometric series which means that we need to factor out of the summation the $j = 0$ term:

$$\begin{aligned} a_{h,k} &= \begin{bmatrix} d-r-2+k-h \\ k-h \end{bmatrix} \\ &\quad \times \sum_{j=0}^{k-h} \frac{(q^{r-d+1};q)_j}{(q;q)_j} \frac{(q^{k-h-j+1};q)_j}{(q^{d-r-2+k-h-j+1};q)_j} q^{j(d-r-1)}. \end{aligned}$$

The second fraction in the summation can be simplified using equation (5.21) from the previous section:

$$(Aq^{-j};q)_j = (-A)^j \, q^{-j(j+1)/2} \, (A^{-1}q;q)_j.$$

We can now write the entry in row h, column k as

$$\begin{aligned} a_{h,k} &= \begin{bmatrix} d-r-2+k-h \\ k-h \end{bmatrix} \\ &\quad \times \sum_{j=0}^{k-h} \frac{(q^{r-d+1};q)_j}{(q;q)_j} \frac{(q^{h-k};q)_j}{(q^{r+2-d+h-k};q)_j} q^{j(d-r-1-d+r+2)} \\ &= \begin{bmatrix} d-r-2+k-h \\ k-h \end{bmatrix} {}_2\phi_1 \left[\begin{matrix} q^{r-d+1}, \ q^{h-k} \\ q^{r+2-d+h-k} \end{matrix} ; q \right]. \end{aligned}$$

We evaluate this summation using equation (5.18), the q-analog of the Chu–Vandermonde formula, with $a = q^{r-d+1}, b = q^{h-k}$, and $c = q^{r+2-d+h-k}$. We see that

$$a_{h,k} = \begin{bmatrix} d-r-2+k-h \\ k-h \end{bmatrix} \frac{(q^{h-k+1};q)_\infty (q^{r+2-d};q)_\infty}{(q^{r+2-d+h-k};q)_\infty (q;q)_\infty}.$$

If k is strictly larger than h, then the first infinite product in the numerator will contain a factor of $1-1=0$. If k equals h, then the infinite products cancel and the Gaussian polynomial is 1. This concludes the proof that the matrix we have called R^{-1} is, indeed, the inverse of R.

Calculation of $T_r^* R^{-1}$

We next calculate the matrix product $T_r^* R^{-1}$, where T_r^* is the matrix T_r with the last row replaced by a row of zeros. We let $b_{h,k}$ denote the entry in row h, column k of this product. Since the last row of T_r^* is all zeroes, so is the last row of $T_r^* R^{-1}$. For $1 \leq h < r$, we have that

$$\begin{aligned} b_{h,k} &= \sum_{j=1}^{r} \lambda q^{h+1-d} \begin{bmatrix} h+j-d \\ j-1 \end{bmatrix} \lambda^{-1} q^{j(d-r-1)} \begin{bmatrix} d-r-2+k-j \\ k-j \end{bmatrix} \\ &= q^{h-r} \sum_{j=0}^{k-1} \begin{bmatrix} h+j+1-d \\ j \end{bmatrix} \begin{bmatrix} d-r-2+k-1-j \\ k-1-j \end{bmatrix} q^{j(d-r-1)}. \end{aligned}$$

We again write this as a basic hypergeometric series by factoring out of the summation the $j=0$ term and then using equation (5.21) to rewrite the second fraction in the summation:

$$b_{h,k} = q^{h-r} \begin{bmatrix} d-r-2+k-1 \\ k-1 \end{bmatrix} \sum_{j=0}^{k-1} \frac{(q^{h+2-d};q)_j \, (q^{1-k};q)_j}{(q;q)_j \, (q^{3+r-d-k};q)_j} \, q^j.$$

We can use equation (5.19) in Corollary 5.3 with $t = k-1$, $b = q^{h+2-d}$, and $c = q^{3+r-d-k}$ to rewrite the sum on j as

$$\frac{(q^{h+1-r};q)_{k-1}}{(q^{d-r-1};q)_{k-1}}.$$

We substitute this into the expression for $b_{h,k}$ and then simplify the product to obtain

$$b_{h,k} = q^{h-r} \frac{(q^{h+1-r};q)_{k-1}}{(q;q)_{k-1}}, \quad 1 \leq h < r.$$

We note that the numerator of this fraction is zero if k is strictly larger than $r-h$, and thus all entries on and below the southwest to northeast diagonal are zero.

Calculation of M^*

Since R is upper triangular, the product $R T_r^* R^{-1}$ will also have all of its nonzero entries above the southwest to northeast diagonal. If $c_{h,k}$ is the entry in row h, column k of this product, then $c_{h,k} = 0$ when $h+k > r$. For $h+k \leq r$, we have that

$$c_{h,k} = \sum_{j=1}^{r} \lambda q^{h(r+1-d)} \begin{bmatrix} r-d+j-h \\ j-h \end{bmatrix} q^{j-r} \frac{(q^{j+1-r};q)_{k-1}}{(q;q)_{k-1}}.$$

These summands are zero unless $h \leq j \leq r - k$. We replace the index of summation j by $j + h$ and then simplify. We get that

$$c_{h,k} = \lambda q^{h(r+1-d)+h-r} \frac{(q^{h+1-r};q)_{k-1}}{(q;q)_{k-1}} \, {}_2\phi_1 \left[\begin{matrix} q^{r-d+1}, \; q^{h+k-r} \\ q^{h+1-r} \end{matrix} ; q \right].$$

We can apply equation (5.19) once again, this time with $t = r - h - k$, $b = q^{r-d+1}$, and $c = q^{h+1-r}$. After simplification, we get get

$$c_{h,k} = (-1)^{k-1} \lambda q^{h(r+k+1-d)-kr+\binom{k}{2}} \begin{bmatrix} 2r-h-d \\ r-h-k \end{bmatrix},$$

provided $r \geq h + k$, while $c_{h,k} = 0$ when $r < h + k$.

The first column of M^* is the rth column of R whose entries happen to equal $-c_{h,0}$, $1 \leq h \leq r$. The ith column of M^* is the $i - 1$st column of $-RA_r^*R^{-1}$. We have therefore shown that

$$M^* = \left((-1)^k \lambda q^{h(r+k+1-d)-kr+\binom{k}{2}} \begin{bmatrix} 2r-h-d \\ r-h-k \end{bmatrix} \right)_{1 \leq h \leq r, \; 0 \leq k \leq r}.$$

The natural extension of this matrix is to the $(r+1) \times (r+1)$ matrix

$$M = \left((-1)^k \lambda q^{h(r+k+1-d)-kr+\binom{k}{2}} \begin{bmatrix} 2r-h-d \\ r-h-k \end{bmatrix} \right)_{h,k=0}^{r}.$$

Finding an eigenvector

We now assume that 1 is an eigenvalue for the matrix M and proceed to look for a corresponding eigenvector. The one clue that we have is that if $\lambda = 1$, $d = 0$, and $q = 1$, then L_k is the number of descending plane partitions in $\mathcal{B}(r+1, r+1, r+1)$ with exactly k parts of size r. From Conjecture 9 on page 24, we expect this number to be

$$L_k = \binom{r+k}{k} \frac{(2r-k)!}{(r-k)!} \prod_{j=0}^{r-1} \frac{(3j+1)!}{(r+j+1)!}.$$

This means that for $\lambda = 1$, $d = 0$, and $q = 1$, the vector whose kth entry is

$$\binom{r+k}{k} \binom{2r-k}{r-k}, \quad 0 \leq k \leq r,$$

should be an eigenvector associated with the eigenvalue 1 for

$$M = \left((-1)^k \binom{2r-h}{r-h-k} \right)_{h,k=0}^{r}.$$

It is not difficult to verify that, in fact, this works.

Now the guessing begins. Since λ is a factor of every term in M, we can ignore it in the initial search for our eigenvector. It is fairly safe to assume that the binomial coefficients should be replaced by Gaussian polynomials and that we will need to multiply each term by some power of q. In view of the way that the parameter d has appeared in our calculations so far, it is also reasonable to assume that we want to subtract d from the numerator parameter in each Gaussian polynomial. Beyond that, we need some flexibility and so, following Mills, Robbins, and Rumsey, we introduce two additional parameters: α and β. In fact, in their original investigations they had introduced many other parameters, eventually discovering that there were only two that were essential. We shall consider the vector $V(\alpha, \beta)$ whose kth entry is

$$V_k(\alpha, \beta) = q^{k(r+1-d+\beta)} \begin{bmatrix} r-d+k \\ k-\alpha \end{bmatrix} \begin{bmatrix} 2r-d-k \\ r-\beta-k \end{bmatrix}, \quad 0 \leq k \leq r.$$

The effect of multiplying this vector on the left by M is to get a new vector,

$$M \cdot V(\alpha, \beta) \;=\; (v_0, v_1, \ldots, v_r)$$

where

$$\begin{aligned} v_h \;&=\; \sum_{k=0}^{r} (-1)^k \lambda q^{h(r+k+1-d)-kr+\binom{k}{2}} \begin{bmatrix} 2r-h-d \\ r-h-k \end{bmatrix} \\ &\qquad \times\, q^{k(r+1-d+\beta)} \begin{bmatrix} r-d+k \\ k-\alpha \end{bmatrix} \begin{bmatrix} 2r-d-k \\ r-\beta-k \end{bmatrix}. \end{aligned}$$

We shall assume that α and β are non-negative integers so that our summands are zero unless $\alpha \leq k \leq r-\beta$. After simplifying the power of q in our summation, we replace the index of summation k by $k+\alpha$, and then factor the $k=0$ term out of the summation:

$$\begin{aligned} v_h \;&=\; (-1)^\alpha \lambda q^{h(r+1-d)+\alpha(h+1-d+\beta)+\binom{\alpha}{2}} \begin{bmatrix} 2r-h-d \\ r-h-\alpha \end{bmatrix} \begin{bmatrix} 2r-d-\alpha \\ r-\alpha-\beta \end{bmatrix} \\ &\qquad \times\, {}_2\phi_1 \left[\begin{matrix} q^{h+\alpha-r},\; q^{\alpha+\beta-r} \\ q^{d+\alpha-2r} \end{matrix} ; q \right]. \end{aligned}$$

Once more, we can sum our series using equation (5.19), this time with $t = r-\alpha-\beta$, $b = q^{h+\alpha-r}$, and $c = q^{d+\alpha-2r}$:

$$\begin{aligned} v_h \;&=\; (-1)^\alpha \lambda q^{h(r+1-d)+\alpha(h+1-d+\beta)+\binom{\alpha}{2}} \begin{bmatrix} 2r-h-d \\ r-h-\alpha \end{bmatrix} \begin{bmatrix} r+h-d \\ h+\alpha+\beta-d \end{bmatrix} \\ &=\; (-1)^\alpha \lambda q^{\alpha(1-d+\beta)+\binom{\alpha}{2}} V_h(d-\alpha-\beta, \alpha). \end{aligned}$$

We have proven that

$$M \cdot V(\alpha,\beta) \;=\; (-1)^{\alpha}\lambda q^{\alpha(1-d+\beta)+\binom{\alpha}{2}}\,V(d-\alpha-\beta,\alpha). \qquad (5.34)$$

If $d=0$ and $\lambda=1$, then $V(0,0)$ is an eigenvector for the eigenvalue 1. The situation is more complicated for $d=2$, $\lambda=q^{1/3}$. We observe that if we define

$$L = V(0,d) + \lambda V(0,0) + \lambda^2 V(d,0),$$

then equation (5.34) implies that

$$M \cdot L = \lambda V(0,0) + \lambda^2 V(d,0) + (-1)^d \lambda^3 q^{-\binom{d}{2}} V(0,d).$$

We see that L will be an eigenvector for the eigenvalue 1 provided $\lambda^3 = (-1)^d q^{\binom{d}{2}}$. When $d=2$ and $\lambda=q^{1/3}$, this equality is satisfied. An eigenvector when $d=2$ and $\lambda=q^{1/3}$ is given by

$$V(0,2) + q^{1/3}V(0,0) + q^{2/3}V(2,0).$$

All that remains is to sum the coordinates of the eigenvector and divide this sum by the first coordinate (corresponding to $k=0$). As we saw in equation (5.32), this will give us the ratio $\det(I_r+T_r)/\det(I_{r-1}+T_{r-1})$. We can verify that our conjectured generating functions are correct if these ratios agree with those given in equations (5.30) and (5.31).

The sum of the coordinates of $V(\alpha,\beta)$, $\sum_{k=0}^{r} V_k(\alpha,\beta)$, equals

$$\sum_{k=\alpha}^{r-\beta} q^{k(r+1-d+\beta)} \begin{bmatrix} r-d+k \\ k-\alpha \end{bmatrix} \begin{bmatrix} 2r-d-k \\ r-\beta-k \end{bmatrix}$$

$$= \quad q^{\alpha(r+1-d+\beta)} \begin{bmatrix} 2r-d-\alpha \\ r-\alpha-\beta \end{bmatrix} {}_2\phi_1 \left[\begin{matrix} q^{r-d+\alpha+1},\ q^{\alpha+\beta-r} \\ q^{d+\alpha-2r} \end{matrix} ;q \right].$$

Yet again, we sum our series using equation (5.19), now with $t=r-\alpha-\beta$, $b=q^{r-d+\alpha+1}$, and $c=q^{d+\alpha-2r}$:

$$\sum_{k=0}^{r} V_k(\alpha,\beta) = q^{\alpha(r+1-d+\beta)} \begin{bmatrix} 3r-2d+1 \\ r-\alpha-\beta \end{bmatrix}.$$

For $d=0$ and $\lambda=1$, the sum of the coordinates of the eigenvector $V(0,0)$ divided by the first coordinate is

$$\frac{\begin{bmatrix} 3r+1 \\ r \end{bmatrix}}{\begin{bmatrix} 2r \\ r \end{bmatrix}} = \prod_{i=1}^{r} \frac{1-q^{2r+1+i}}{1-q^{r+i}},$$

which proves Andrews's conjecture.

For $d = 2$ and $\lambda = q^{1/3}$, the eigenvector is

$$V(0,2) + q^{1/3}V(0,0) + q^{2/3}V(2,0).$$

The sum of the coordinates of this vector is

$$\begin{bmatrix} 3r-3 \\ r-2 \end{bmatrix} + q^{1/3}\begin{bmatrix} 3r-3 \\ r \end{bmatrix} + q^{2(r-1)+2/3}\begin{bmatrix} 3r-3 \\ r-2 \end{bmatrix}$$

$$= \frac{(q^{2r};q)_{r-2}}{(q;q)_r}(1-q^{r-1})(1+q^{1/3})(1-q^{r-1/3})\,\frac{1-q^{3r-2}}{1-q^{r-2/3}}.$$

The first coordinate of this vector is $V_0(0,2) + q^{1/3}V_0(0,0) + q^{2/3}V_0(2,0)$ which is equal to

$$\begin{bmatrix} 2r-2 \\ r-2 \end{bmatrix} + q^{1/3}\begin{bmatrix} 2r-2 \\ r \end{bmatrix} = \frac{(q^{r-1};q)_r}{(q;q)_r}(1+q^{1/3}).$$

The sum of the coordinates of the eigenvector divided by the first coordinate is

$$\frac{(q^{2r};q)_{r-2}\,(1-q^{r-1})(1-q^{r-1/3})(1-q^{3r-2})}{(q^{r-1};q)_r}$$

$$= \frac{1-q^{r-1/3}}{1-q^{r-2/3}}\prod_{i=1}^{r-1}\frac{1-q^{2r-1+i}}{1-q^{r-1+i}}.$$

This proves Macdonald's conjecture.

Q.E.D.

We have reached our second summit, but by a route no one had anticipated. Those who had been working with symmetric functions and the theory of partitions realized that there was something deep and important lying under the alternating sign matrix conjecture. Work on this conjecture began to move into the mainstream of algebraic combinatorics.

Exercises

5.3.1 Prove that

$$\prod_{\eta\in\mathcal{B}(r,r,r)/\mathcal{C}_3}\frac{1-q^{|\eta|+\mathrm{ht}(\eta)}}{1-q^{\mathrm{ht}(\eta)}}\Bigg/\prod_{\eta\in\mathcal{B}(r-1,r-1,r-1)/\mathcal{C}_3}\frac{1-q^{|\eta|+\mathrm{ht}(\eta)}}{1-q^{\mathrm{ht}(\eta)}}$$

$$= \frac{1-q^{3r-1}}{1-q^{3r-2}}\prod_{i=1}^{r-1}\frac{1-q^{6r-3+3i}}{1-q^{3r-3+3i}}.$$

Hint: The product on the left is the product over all orbits in $\mathcal{B}(r,r,r)$ with at least one coordinate equal to r. One of these orbits contains a single point: (r,r,r). Each of the other orbits contains exactly one point of the form (i,j,r) where $1 \le i \le r-1$ and $1 \le j \le r$.

5.3.2 Verify that equations (5.28) and (5.29) are correct for $r=1$ and $r=2$.

5.3.3 Prove that

$$\left(\binom{r}{0}\binom{2r}{r}, \ldots, \binom{r+j}{j}\binom{2r-j}{r-j}, \ldots, \binom{2r}{r}\binom{r}{0}\right)$$

is an eigenvector, with eigenvalue 1, for

$$M = \left((-1)^k \binom{2r-h}{r-h-k}\right)_{h,k=0}^{r}.$$

5.3.4 Prove that

$$\prod_{1\le i\le j\le r+1} \frac{1-q^{r+i+j}}{1-q^{2i+j-1}} \Big/ \prod_{1\le i\le j\le r} \frac{1-q^{r+i+j-1}}{1-q^{2i+j-1}} = \prod_{i=1}^{r} \frac{1-q^{2r+i+1}}{1-q^{r+i}}.$$

5.3.5 Prove that

$$\sum_{j=1}^{r} \lambda q^{h(r+1-d)} \begin{bmatrix} r-d+j-h \\ j-h \end{bmatrix} q^{j-r} \frac{(q^{j+1-r};q)_{k-1}}{(q;q)_{k-1}}$$

$$= \lambda q^{h(r+1-d)+h-r} \frac{(q^{h+1-r};q)_{k-1}}{(q;q)_{k-1}} \, {}_2\phi_1 \left[\begin{matrix} q^{r-d+1}, \; q^{h+k-r} \\ q^{h+1-r} \end{matrix} ; q \right].$$

5.3.6 Let L, M, and N be square matrices that are identical except for their bottom rows. Let R be the bottom row of L, S be the bottom row of M, and $R+S$ (the vector sum) be the bottom row of N. Prove that $\det(L)+\det(M)=\det(N)$.

5.3.7 Prove that

$$\lambda q^{h(r+1-d)+h-r} \frac{(q^{h+1-r};q)_{k-1}}{(q;q)_{k-1}} \, {}_2\phi_1 \left[\begin{matrix} q^{r-d+1}, \; q^{h+k-r} \\ q^{h+1-r} \end{matrix} ; q \right]$$

$$= (-1)^{k-1} \lambda q^{h(r+k+1-d)-kr+\binom{k}{2}} \begin{bmatrix} 2r-h-d \\ r-h-k \end{bmatrix}.$$

Note that this requires applying equation (5.19) and then using equation (5.21) to reverse one of the rising q-factorials.

5.3.8 Prove that

$$\sum_{k=0}^{r}(-1)^k \lambda q^{h(r+k+1-d)-kr+\binom{k}{2}} \begin{bmatrix} 2r-h-d \\ r-h-k \end{bmatrix} \times q^{k(r+1-d+\beta)} \begin{bmatrix} r-d+k \\ k-\alpha \end{bmatrix} \begin{bmatrix} 2r-d-k \\ r-\beta-k \end{bmatrix}$$

$$= (-1)^\alpha \lambda q^{h(r+1-d)+\alpha(h+1-d+\beta)+\binom{\alpha}{2}} \begin{bmatrix} 2r-h-d \\ r-h-\alpha \end{bmatrix} \times \begin{bmatrix} 2r-d-\alpha \\ r-\alpha-\beta \end{bmatrix} {}_2\phi_1 \left[\begin{matrix} q^{h+\alpha-r},\ q^{\alpha+\beta-r} \\ q^{d+\alpha-2r} \end{matrix} ; q \right].$$

5.3.9 Prove that

$$(-1)^\alpha \lambda q^{h(r+1-d)+\alpha(h+1-d+\beta)+\binom{\alpha}{2}} \begin{bmatrix} 2r-h-d \\ r-h-\alpha \end{bmatrix} \begin{bmatrix} 2r-d-\alpha \\ r-\alpha-\beta \end{bmatrix} \times {}_2\phi_1 \left[\begin{matrix} q^{h+\alpha-r},\ q^{\alpha+\beta-r} \\ q^{d+\alpha-2r} \end{matrix} ; q \right]$$

$$= (-1)^\alpha \lambda q^{h(r+1-d)+\alpha(h+1-d+\beta)+\binom{\alpha}{2}} \times \begin{bmatrix} 2r-h-d \\ r-h-\alpha \end{bmatrix} \begin{bmatrix} r+h-d \\ h+\alpha+\beta-d \end{bmatrix}.$$

Again, this simplification requires equation (5.19).

5.3.10 Prove that

$$\sum_{k=\alpha}^{r-\beta} q^{k(r+1-d+\beta)} \begin{bmatrix} r-d+k \\ k-\alpha \end{bmatrix} \begin{bmatrix} 2r-d-k \\ r-\beta-k \end{bmatrix}$$

$$= q^{\alpha(r+1-d+\beta)} \begin{bmatrix} 2r-d-\alpha \\ r-\alpha-\beta \end{bmatrix} {}_2\phi_1 \left[\begin{matrix} q^{r-d+\alpha+1},\ q^{\alpha+\beta-r} \\ q^{d+\alpha-2r} \end{matrix} ; q \right].$$

5.3.11 Prove that

$$q^{\alpha(r+1-d+\beta)} \begin{bmatrix} 2r-d-\alpha \\ r-\alpha-\beta \end{bmatrix} {}_2\phi_1 \left[\begin{matrix} q^{r-d+\alpha+1},\ q^{\alpha+\beta-r} \\ q^{d+\alpha-2r} \end{matrix} ; q \right]$$

$$= q^{\alpha(r+1-d+\beta)} \begin{bmatrix} 3r-2d+1 \\ r-\alpha-\beta \end{bmatrix}.$$

5.3.12 Verify that

$$\begin{bmatrix} 3r-3 \\ r-2 \end{bmatrix} + q^{1/3} \begin{bmatrix} 3r-3 \\ r \end{bmatrix} + q^{2(r-1)+2/3} \begin{bmatrix} 3r-3 \\ r-2 \end{bmatrix}$$

$$= \frac{(q^{2r};q)_{r-2}}{(q;q)_r}(1-q^{r-1})(1+q^{1/3})(1-q^{r-1/3})\,\frac{1-q^{3r-2}}{1-q^{r-2/3}}.$$

Much more serious than not knowing whether a given fact is true, is the agony of realizing that our cherished *tools of the trade* are inadequate to tackle a given problem. The fact that a conjecture resists vigorous attacks by skilled practitioners is an impetus for us either to sharpen our existing tools, or else create new ones. The value of a proof of an outstanding conjecture should be judged, not by its cleverness and elegance, and not even by its "explanatory power," but by the extent in which it enlarges our toolbox. By this standard, the present proof is adequate. Like most new tools, the present method of proof is a judicious assembly of existing tools, which I will now describe.

The first ingredient consists of *partial recurrence equations* (alias *partial difference equations*) and operators. The calculus of finite differences was introduced in the last century by discrete mathematician George Boole, but in this century was taken up, and almost monopolized by, continuous number crunchers, who called them *finite difference schemes*. A notable exception was Dick Duffin (e.g. Duffin 1956) through whose writings I learnt about these objects, and immediately fell in love with them (e.g. Zeilberger 1980a, 1980b). It was fun returning to my first love.†

Conspicuously missing from the present paper is my second love, *bijective proofs* (e.g. Zeilberger and Bressoud 1985) that were taught to me by Dominique Foata, Xavier G. Viennot, Herb Wilf and many others. However, doing bijections made me a better mathematician and person, so their implicit impact is considerable.

The second ingredient is my third love, *constant term* identities introduced to me by Dick Askey. Dennis Stanton (1986) and John Stembridge (1988) showed me how to crack them (Zeilberger 1988, 1990). The *Stanton-Stembridge trick* was indeed crucial.

The third and last ingredient, which is not mentioned explicitly, but without which this proof could never have come to be, is my current love: computer algebra and Maple. Practically every lemma, sublemma, subsublemma ..., was first conjectured with the aid of Maple, and then tested by it. A Maple package, `ROBBINS`, that empirically (and in a few cases rigorously) checks every non-trivial fact proved in this paper, is given as a companion to this paper, and should be used in conjunction with it.

– Doron Zeilberger, from the Introduction to
Proof of the alternating sign matrix conjecture (1996a)

† "Mathematicians, like Proust and everyone else, are at their best when writing about their first love." – Gian-Carlo Rota (Kac, Rota, and Schwartz 1992, p. 3)

6
Explorations

These conjectures are of such compelling simplicity that it is hard to understand how any mathematician can bear the pain of living without understanding why they are true.

– David Robbins (1991)

If there had been any lingering doubt of the truth of Conjectures 1 and 2, it was dispelled by the proof we have just seen of Macdonald's conjecture. The critical insight that made this proof possible came from the assumption that Conjectures 1 and 2 were correct. No further confirmation was needed, but there was still no proof.

6.1 Charting the territory

The alternating sign matrix conjecture appeared at an opportune moment. The publication of George Andrews's *The Theory of Partitions* (1976) and Ian Macdonald's *Symmetric Functions and Hall Polynomials* (1979) had created wide interest in plane partitions and Young tableaux. Enough people knew enough about them to recognize the potential importance of proving this conjecture. Because these books could take researchers from disparate backgrounds and bring them quickly up to the edge of what was known, they drew in mathematicians from a variety of fields, including algebra, combinatorics, and analysis. More important, even though it would be ten years before any significant progress would be made on the proof of this conjecture, those who were drawn to it found that it was only one in a constellation of related problems, and that there were many other interesting discoveries to be made. The work of Gessel and Viennot on the counting of non-intersecting lattice paths followed the Mills, Robbins, and Rumsey proof and was largely motivated by the counting problems that appeared around the alternating sign matrix conjecture.

To anyone new to this conjecture, the initial response was to try to find a one-to-one correspondence between $n \times n$ alternating sign matrices and descending plane partitions with largest part less than or equal to n. As mentioned in Chapter 1, the first thing that strikes anyone looking for such a correspondence is that the set of alternating sign matrices is rich in symmetries and natural parameters. The descending plane partitions have no obvious symmetries and few obvious parameters. If there is a natural one-to-one correspondence, then there must be hidden symmetries in the set of descending plane partitions. The first of these that Mills, Robbins, and Rumsey surmised was that the number of descending plane partitions with j parts of size n must equal the number of descending plane partitions with $n-1-j$ parts of size n. It is not difficult to see why this is true for $j=0$ (see page 24). They were able to prove that these numbers are the same, but they could not find a simple explanation of why this should be so.

In the mid-1980s, Mills, Robbins, and Rumsey published two papers (1983, 1986) of insights into and conjectures about the correspondence between alternating sign matrices and descending plane partitions. There are three results in these papers that are significant for our story. The first of these addresses the question of what parameters for descending plane partitions should correspond to the number of -1s in the alternating sign matrix and to its inversion number (definition on page 88). Given a descending plane partition,

$$\begin{array}{cccccccc} a_{1,1} & a_{1,2} & a_{1,3} & \cdots & \cdots & \cdots & \cdots & a_{1,r_1} \\ & a_{2,2} & a_{2,3} & \cdots & \cdots & \cdots & a_{2,r_2} & \\ & & & & \vdots & & & \\ & & & a_{k,k} & \cdots & a_{k,r_k}, & & \end{array}$$

a **special part** is an entry that satisfies $a_{i,j} \le j-i$. Mills, Robbins, and Rumsey conjectured that the number of special parts in a descending plane partition should correspond to the number of -1s in the matching alternating sign matrix and that the number of parts in the descending plane partition should correspond to the inversion number of the alternating sign matrix. Specifically, they made the following conjecture which is still unproven, though it has been verified for all values of n less than or equal to 7 and for $m=0$.

Conjecture 10 *Let $A(n,k,m,p)$ be the number of $n \times n$ alternating sign matrices with a 1 in the kth column of the first row, with m -1s, and with inversion number equal to p. Let $D(n,k,m,p)$ be the number of descending plane partitions with largest part less than or equal to n, with exactly $k-1$ parts of size n, with m special parts, and with a total of p parts. We then have that*

$$A(n,k,m,p) = D(n,k,m,p). \tag{6.1}$$

The second result that concerns us involves the polynomial

$$f_n(x) = \sum_{A \in \mathcal{A}_n} x^{N(A)},$$

where $\mathcal{A}_n$ is the set of $n \times n$ alternating sign matrices and $N(A)$ is the number of -1s in A. The recursive algorithm described in exercise 3.2.10 on page 91 can be specialized (set $q = 1$) to calculate the polynomial $f_n(x)$. This yields:

$$\begin{aligned}
f_1(x) &= 1, \\
f_2(x) &= 2, \\
f_3(x) &= 6 + x, \\
f_4(x) &= 24 + 16x + 2x^2, \\
f_5(x) &= 120 + 200x + 94x^2 + 14x^3 + x^4, \\
f_6(x) &= 720 + 2400x + 2684x^2 + 1284x^3 + 310x^4 + 36x^5 + 2x^6, \\
f_7(x) &= 5040 + 24900x + 63308x^2 + 66158x^3 + 38390x^4 + 13037x^5 \\
&\quad + 2660x^6 + 328x^7 + 26x^8 + x^9.
\end{aligned}$$

It follows from the definition of this function that $f_n(0) = n!$ and $f_n(1)$ is the number of $n \times n$ alternating sign matrices. As we saw in Section 3.5,

$$\sum_{B \in \mathcal{A}_n} \lambda^{\mathcal{I}(B)} (1 + \lambda^{-1})^{N(B)} = (1 + \lambda)^{n(n-1)/2}.$$

When we set λ equal to 1, we get that $f_n(2) = 2^{n(n-1)/2}$. Mills, Robbins, and Rumsey conjectured that there would also be an appealing formula for $f_n(3)$.

Conjecture 11 *For all positive integers n we have that*

$$\frac{f_{2n+1}(3)}{f_{2n}(3)} = 3^n \binom{3n}{n} \Big/ \binom{2n}{n}, \tag{6.2}$$

$$\frac{f_{2n}(3)}{f_{2n-1}(3)} = 3^{n-1}\binom{3n-1}{n} \Big/ \binom{2n-1}{n}. \qquad (6.3)$$

Greg Kuperberg proved this conjecture in 1995 (Kuperberg 1996b†). It was a serendipitous consequence of his proof of Conjecture 3 using the square ice model. More than this, Kuperberg was able to find a general formula for $f_n(z)$. This formula can be rewritten as a product when $x^2 + (2 - z)x + 1$ is a cyclotomic polynomial in x. That is to say, there is a product formula when $z = 1$, 2, or 3.

A hidden symmetry

For the purposes of our story, however, the most important result of the 1983 paper was an insight into the hidden symmetry of descending plane partitions that corresponds to the reflection of an alternating sign matrix over a vertical axis. This "reflection" takes a descending plane partition with $k-1$ parts of size n to a uniquely defined companion with $n - k$ parts of size n. It preserves the number of special parts (which should correspond to the number of -1s in the alternating sign matrix). If m is the number of special parts and p is the total number of parts, then this reflection of a descending plane partition changes the total number of parts to $n(n-1)/2 + m - p$, which is precisely what happens to the inversion number when an alternating sign matrix is reflected over a vertical axis.

The mapping that accomplishes this first takes the descending plane partition with entries a_{ij}, $j \geq i$, to a complementary array, $b_{ij}, 1 \leq i \leq j \leq n-1$, defined by

$$b_{ij} = \begin{cases} j - i + 1 - a_{ij}, & \text{if } a_{ij} \text{ exists and } a_{ij} \leq j - i, \\ j + 1 - \beta_{ij}, & \text{if } a_{ij} \text{ is not defined,} \\ \text{undefined} & \text{if } a_{ij} > j - i, \end{cases}$$

where β_{ij} is the number of "oversized" entries in column i. Specifically, β_{ij} is the cardinality of $\{a_{xi} \mid a_{xi} \geq j + 2 - x\}$. One example of such a

† The initially published proof is not correct. It has since been corrected.

corresponding pair of arrays is given by

$$\begin{array}{cccccc} 7 & 6 & 6 & 5 & 4 & 4 \\ & 5 & 5 & 3 & 3 & - \\ & & 3 & 2 & - & - \\ & & & - & - & - \\ & & & & - & - \\ & & & & & - \end{array} \quad \Longleftrightarrow \quad \begin{array}{cccccc} - & - & - & - & 1 & 2 \\ & - & - & - & 1 & 7 \\ & & - & - & 4 & 7 \\ & & & 4 & 6 & 7 \\ & & & & 6 & 7 \\ & & & & & 7 \end{array}.$$

The second step of our transformation is to take the mirror image of the new array across the southwest to northeast diagonal,

$$\begin{array}{cccccc} 7 & 6 & 6 & 5 & 4 & 4 \\ & 5 & 5 & 3 & 3 & - \\ & & 3 & 2 & - & - \\ & & & - & - & - \\ & & & & - & - \\ & & & & & - \end{array} \quad \Longleftrightarrow \quad \begin{array}{cccccc} 7 & 7 & 7 & 7 & 7 & 2 \\ & 6 & 6 & 4 & 1 & 1 \\ & & 4 & - & - & - \\ & & & - & - & - \\ & & & & - & - \\ & & & & & - \end{array}.$$

Having found a natural involution on the set of descending plane partitions, the next step was to ask what its analog might be for cyclically symmetric plane partitions. They soon discovered that it would be simple complementation: Take the box in which the cyclically symmetric plane partition sits, and consider the plane partition formed by the cubes that are not in that plane partition. This led Robbins to start thinking about complements of arbitrary plane partitions. In particular, he asked the question: How many plane partitions inside a box of given dimensions are left unchanged when one takes the complement? In his own words (personal communication), this is what happened:

> One evening while I was in upstate Michigan on vacation, it occurred to me that the plane partitions with the maximum symmetry would be totally symmetric self-complementary plane partitions. At first I thought perhaps such things could not exist, but before long I was counting them and realized that there was 1 way to do it in a 2x2x2 box, 2 ways to do it in a 4x4x4 box and 7 ways to do it in a 6x6x6 box ... It was quite a magical experience.

David Robbins had found another conjecture that would eventually (Andrews 1994) be proven to be equivalent to the alternating sign matrix conjecture.

Conjecture 12 *The number of totally symmetric self-complementary plane partitions that fit inside $\mathcal{B}(2n, 2n, 2n)$ (see Fig. 6.1) is equal to the number of $n \times n$ alternating sign matrices.*

Figure 6.1. The seven totally symmetric self-complementary plane partitions in $\mathcal{B}(6,6,6)$.

Robbins shared his insight with Richard Stanley, who realized that this suggested other symmetry conditions on plane partitions.

Symmetries of plane partitions

Given a box $\mathcal{B}(r,s,t)$ and a plane partition π that fits inside it, we define the **complement** of π, π^c, to be the reflection across the $y=x$ plane of the set of lattice points that are in $\mathcal{B}(r,s,t)$ but not in π:

$$\pi^c = \{(i,j,k) \in \mathcal{B}(r,s,t) \mid (r-i-1, s-j-1, t-k-1) \notin \pi\}. \quad (6.4)$$

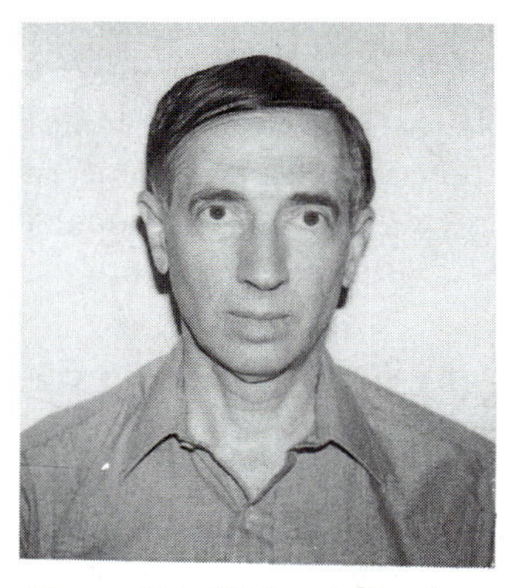

Figure 6.2. Richard Stanley.

This introduces a third transformation of a plane partition. MacMahon had studied plane partitions that are invariant under reflection across the $y = x$ plane: symmetric plane partitions. Macdonald had introduced plane partitions that are invariant under cyclic rotations of the axes – cyclically symmetric plane partitions – or under the full group of permutations of the axes – totally symmetric plane partitions. With the introduction of complementation, there are six new invariances that can be studied.

For the sake of completeness, we begin with the four counting functions that by now had become classical. The first three of these are special cases of the generating functions for which we have seen proofs. In the following formulæ, we let D_r be the number of descending plane partitions with largest part less than or equal to r,

$$D_r = \prod_{j=0}^{r-1} \frac{(3j+1)!}{(r+j)!}.$$

1. **Plane partitions without symmetry conditions.** We let $N_1(r,s,t)$ denote the total number of plane partitions in $\mathcal{B}(r,s,t)$. This is the $q = 1$ case of the generating function identity that was first stated and proven by MacMahon and given in this book as Theorem 1.3. We have seen one proof in Chapter 3 and a second proof in Chapter 4:

$$N_1(r,s,t) = \prod_{(i,j,k)\in\mathcal{B}(r,s,t)} \frac{i+j+k-1}{i+j+k-2} = \prod_{i=1}^{r}\prod_{j=1}^{s} \frac{i+j+t-1}{i+j-1}. \tag{6.5}$$

2. **Symmetric plane partitions.** Let $N_2(r,s,t)$ be the number of plane partitions in $\mathcal{B}(r,s,t)$ that are symmetric with respect to the $y=x$ plane. This counting function is the $q=1$ case of the generating function that was conjectured by MacMahon and proven by Andrews (1978) and Macdonald (1979). We have seen a proof of it in Chapter 4:

$$N_2(r,r,s) = \left(\prod_{i=1}^{r} \frac{2i+s-1}{2i-1}\right) \left(\prod_{1\le i<j\le r} \frac{i+j+s-1}{i+j-1}\right). \quad (6.6)$$

3. **Cyclically symmetric plane partitions.** Let $N_3(r,s,t)$ be the number of cyclically symmetric plane partitions in $\mathcal{B}(r,s,t)$. Macdonald (1979) conjectured the generating function for which this is the $q=1$ case. This particular formula was first proven by Andrews (1979) as the "weak" Macdonald conjecture. The full Macdonald conjecture was proven by Mills, Robbins, and Rumsey in 1982. We have seen a proof of it in Chapter 5:

$$N_3(r,r,r) = \left(\prod_{i=1}^{r} \frac{2i+r-1}{3i-2}\right) \left(\prod_{1\le i<j\le r} \frac{i+j+r-1}{i+j-1}\right). \quad (6.7)$$

4. **Totally symmetric plane partitions.** Let $N_4(r,s,t)$ be the number of totally symmetric plane partitions in $\mathcal{B}(r,s,t)$. Conjectured by Macdonald, Stembridge (1995) proved this formula:

$$N_4(r,r,r) = \prod_{1\le i\le j\le r} \frac{i+j+r-1}{i+j-1}. \quad (6.8)$$

5. **Self-complementary plane partitions.** Let $N_5(r,s,t)$ be the number of self-complementary plane partitions in $\mathcal{B}(r,s,t)$. The product rst must be even. Richard Stanley (1986b) proved that

$$N_5(2r,2s,2t) = N_1(r,s,t)^2, \quad (6.9)$$

$$N_5(2r+1,2s,2t) = N_1(r,s,t)\,N_1(r+1,s,t), \quad (6.10)$$

$$N_5(2r+1,2s+1,2r) = N_1(r+1,s,t)\,N_1(r,s+1,t). \quad (6.11)$$

6. **Transpose complement plane partitions.** These are plane partitions for which the complement is the same as the reflection in the $y=x$ plane. Let $N_6(r,s,t)$ be the number of such plane partitions in $\mathcal{B}(r,s,t)$. The parameters r and s must be equal

and t must be even. Robert Proctor (1988) proved that

$$N_6(r, r, 2t) = \binom{t+r-1}{r-1} \prod_{1 \le i \le j \le r-2} \frac{2t+i+j+1}{i+j+1}. \quad (6.12)$$

7. **Symmetric self-complementary plane partitions.** We let $N_7(r, s, t)$ be the number of such plane partitions in $\mathcal{B}(r, s, t)$. The parameters r and s must be equal and t must be even. Proctor (1983) proved that

$$N_7(2r, 2r, 2t) = N_1(r, r, t), \quad (6.13)$$
$$N_7(2r+1, 2r+1, 2t) = N_1(r, r+1, t). \quad (6.14)$$

8. **Cyclically symmetric transpose complement plane partitions.** Let $N_8(r, s, t)$ be the number of such plane partitions in $\mathcal{B}(r, s, t)$. The parameters r, s, and t must be equal and even. Mills, Robbins, and Rumsey (1983) proved that

$$N_8(2r, 2r, 2r) = \prod_{i=0}^{r-1} \frac{(3i+1)!(6i)!(2i)!}{(4i+1)!(4i)!}. \quad (6.15)$$

9. **Cyclically symmetric self-complementary plane partitions.** Let $N_9(r, s, t)$ be the number of such plane partitions in $\mathcal{B}(r, s, t)$. The parameters r, s, and t must be equal and even. Kuperberg (1994) proved that

$$N_9(2r, 2r, 2r) = D_r^2. \quad (6.16)$$

10. **Totally symmetric self-complementary plane partitions.** Let $N_{10}(r, s, t)$ be the number of such plane partitions in $\mathcal{B}(r, s, t)$. The parameters r, s, and t must be equal and even. This was the big surprise, first observed by David Robbins and finally proven by George Andrews (1994),

$$N_{10}(2r, 2r, 2r) = D_r. \quad (6.17)$$

Equation (6.17) implies that the number of $n \times n$ alternating sign matrices, A_n, should be the same as the number of totally symmetric self-complementary plane partitions in $\mathcal{B}(2n, 2n, 2n)$. In Section 3 of this chapter, we shall see that the first proof of Conjecture 3, the formula for A_n, was accomplished by proving that $A_n = N_{10}(2n, 2n, 2n)$ and then invoking Andrews's proof that $N_{10}(2n, 2n, 2n) = D_n$. But the proof of equation (6.17) has a story of its own that I want to tell first. We shall see it in the next section.

The −1 phenomenon

In 1994, John Stembridge published a proof of an observation that he had made about these ten formulæ. It begins with the recognition that there are two natural ways of forming the generating function for plane partitions: We can let the power of q count the total number of unit cubes in the partitions (the way we have been defining our generating functions), or, given a subgroup G of $\mathcal{S}_3$ under which the plane partitions are invariant, we could let the power of q count the number of G-orbits in the plane partition.

As we have seen, the generating function that counts the number of unit cubes is given by

$$\prod_{\eta \in \mathcal{B}/G} \frac{1 - q^{|\eta|(1+\mathrm{ht}(\eta))}}{1 - q^{|\eta|\,\mathrm{ht}(\eta)}},$$

provided G is I (the identity), $\mathcal{S}_2$ (the identity and one transposition of two of the axes), or C_3 (all cyclic permutations of the axes). While it is not correct when $G = S_3$, the $q = 1$ case does give the correct number of totally symmetric plane partitions inside $\mathcal{B}$. Whether or not it is correct, we shall call this the **expected formula** for the unit cube counting generating function.

There is also an expected formula for the generating function that counts the number of orbits:

$$\prod_{\eta \in \mathcal{B}/G} \frac{1 - q^{1+\mathrm{ht}(\eta)}}{1 - q^{\mathrm{ht}(\eta)}}.$$

This formula is correct when $G = I$ or S_2. It is wrong when $G = C_3$, although the $q = 1$ case does give the correct count for the total number of cyclically symmetric plane partitions inside $\mathcal{B}$. The fact that the orbit counting generating function for S_2 is given by the expected formula is equivalent to the Bender–Knuth conjecture (see exercise 4.3.7 on page 147). The case $G = \mathcal{S}_3$ is a conjecture that George Andrews and David Robbins arrived at independently and that is still unproven.

Conjecture 13 *The orbit counting generating function for totally symmetric plane partitions that fit inside $\mathcal{B}$ is given by the expected formula:*

$$\prod_{\eta \in \mathcal{B}/\mathcal{S}_3} \frac{1 - q^{1+\mathrm{ht}(\eta)}}{1 - q^{\mathrm{ht}(\eta)}}.$$

What Stembridge observed is that for any subgroup G, the number of plane partitions invariant under G and complementation (N_5, N_7, N_9, or N_{10}) is equal to the limit as q approaches -1 of the orbit counting generating function for plane partitions invariant under G. The number of plane partitions invariant under $G = \mathcal{S}_2$ or C_3 and whose transpose corresponds to the complement (N_6 or N_8) is equal to the limit as q approaches -1 of the unit cube counting generating function for plane partitions invariant under G. A curious feature of this observation is that even when the corresponding generating function is not given by the expected formula, both the true generating function evaluated at $q = -1$ and the expected generating function evaluated at $q = -1$ produce the same number. While Stembridge has proven that this is so, no satisfactory explanation of this phenomenon has been found.

More conjectures

Many more conjectures have been made. Some are treated in the exercises to this section. I mention just a few more here. We can count alternating sign matrices that satisfy certain symmetry conditions and ask when these numbers are **smooth**, that is to say, when they have only small prime divisors. In such cases, there is likely to be a formula as a rational product of factorials. Let VS(n), HTS(n), QTS(n), DDS(n), and VHS(n) (respectively) be the number of $n \times n$ alternating sign matrices that are left unchanged by (respectively) vertical reflection, half-turn or 180° rotation, quarter-turn or 90° rotation, reflection across either diagonal, and vertical or horizontal reflection. Conjectured and still unproven formulæ for these counting functions are

$$\mathrm{VS}(2n+1)/\mathrm{VS}(2n-1) = \binom{6n-2}{2n} \Big/ 2\binom{4n-1}{2n}, \tag{6.18}$$

$$\mathrm{HTS}(2n+1)/\mathrm{HTS}(2n) = \binom{3n}{n} \Big/ \binom{2n}{n}, \tag{6.19}$$

$$\mathrm{HTS}(2n)/\mathrm{HTS}(2n-1) = 4\binom{3n}{n} \Big/ 3\binom{2n}{n}, \tag{6.20}$$

$$\begin{aligned} \mathrm{QTS}(4n) &= \mathrm{HTS}(2n) D_n^2 \\ &= N_3(n,n,n) D_n^3, \end{aligned} \tag{6.21}$$

$$\mathrm{QTS}(4n+1) = \mathrm{HTS}(2n+1) D_n^2, \tag{6.22}$$

$$\mathrm{QTS}(4n-1) = \mathrm{HTS}(2n-1) D_n^2, \tag{6.23}$$

$$\mathrm{DDS}(2n+1)/\mathrm{DDS}(2n-1) = \binom{3n}{n} \Big/ \binom{2n-1}{n}, \qquad (6.24)$$

$$\mathrm{VHS}(4n+1)/\mathrm{VHS}(4n-1) = \frac{3n-1}{4n-1}\binom{6n-3}{2n-1} \Big/ \binom{4n-2}{2n-1}, \qquad (6.25)$$

$$\mathrm{VHS}(4n+3)/\mathrm{VHS}(4n+1) = \frac{3n+1}{4n+1}\binom{6n}{2n} \Big/ \binom{4n}{2n}. \qquad (6.26)$$

Exercises

6.1.1 Prove that if $x^2 + (2 - z)x + 1$ is a cyclotomic polynomial in x, then $z = 1$, 2, or 3.

6.1.2 Prove the following special case of Conjecture 10: The number of permutations $\sigma \in \mathcal{S}_n$ for which $\sigma(1) = k$ and $\mathcal{I}(\sigma) = p$ is equal to the number of descending plane partitions in $\mathcal{B}(n,n,n)$ with exactly $k-1$ parts of size n, no special parts (the entry in position (i,j) must be strictly greater than $j-i$), and a total of p parts.

6.1.3 The *Mathematica* function `MTf[{j},n,x,q]` defined on page 91, is the sum of $q^{\mathcal{I}(B)}x^{N(B)}$ taken over all $n \times n$ alternating sign matrices B that have a 1 in the first row and jth column. Let $A_{n,k}(x) =$ `MTf[{k},n,x,1]` and let $p_n(x)$ be the greatest common divisor of $\{A_{n,1}(x), A_{n,2}(x), \ldots, A_{n,n}(x)\}$. Mills, Robbins, and Rumsey (1983) conjectured that if n is odd, then `ASMf[n,x,1]` equals $p_n(x)p_{n+1}(x)$, while if n is even, then `ASMf[n,x,1]` equals $2p_n(x)p_{n+1}(x)$. Check this conjecture as far as you can.

6.1.4 Another conjecture of Mills, Robbins, and Rumsey is that when n is odd, $p_n(x)$ equals `MTf[{1,3,5,...,n},n,x,1]`. Check this conjecture as far as you can.

6.1.5 Prove that if an $n \times n$ alternating sign matrix with m -1s and inversion number equal to p is reflected over a vertical axis, it is transformed into an alternating sign matrix whose inversion number is $n(n-1)/2 + m - p$.

6.1.6 Prove that the image of a descending plane partition under the reflection described on page 194 is always a descending plane partition.

6.1.7 Prove that the reflection of a descending plane partition takes a descending plane partition in which n appears j times to a descending plane partition in which n appears $n-1-j$ times.

6.1.8 Prove that applying the Mills, Robbins, and Rumsey reflection to a descending plane partition twice returns the original descending plane partition.

6.1.9 Prove that the orbit counting generating function for symmetric plane partitions that fit inside $\mathcal{B}(r,r,t)$ is the same as the unit cube counting generating function for column strict plane partitions with at most r rows, t columns, and largest part less than or equal to r. Verify that

$$\prod_{1 \le i \le j \le r} \frac{1-q^{t+i+j-1}}{1-q^{i+j-1}} = \prod_{\eta \in \mathcal{B}(r,r,t)/\mathcal{S}_2} \frac{1-q^{1+\mathrm{ht}(\eta)}}{1-q^{\mathrm{ht}(\eta)}}.$$

6.1.10 Prove that

$$\lim_{q \to -1} \prod_{1 \le i \le j \le k \le n} \frac{1-q^{i+j+k-1}}{1-q^{i+j+k-2}} = \prod_{j=0}^{n-1} \frac{(3j+1)!}{n+j)!}.$$

6.1.11 Verify that

$$\lim_{q \to -1} \prod_{\eta \in \mathcal{B}/G} \frac{1-q^{1+\mathrm{ht}(\eta)}}{1-q^{\mathrm{ht}(\eta)}}$$

for $G = I$, $\mathcal{S}_2$, or C_3 yields the product formulas for N_5, N_7, or N_9, respectively.

6.1.12 Verify that

$$\lim_{q \to -1} \prod_{\eta \in \mathcal{B}/G} \frac{1-q^{|\eta|(1+\mathrm{ht}(\eta))}}{1-q^{|\eta|\,\mathrm{ht}(\eta)}}$$

for $G = \mathcal{S}_2$ or C_3 yields the product formulas for N_6 or N_8, respectively.

6.2 Totally symmetric self-complementary plane partitions

The story of the proof of the enumeration formula for totally symmetric self-complementary plane partitions (which hereafter we shall refer to as **TSSCPPs**) is a tale of a sequence of mathematicians who each took the problem one step forward, and then passed it on to someone else. It begins with Mills, Robbins, and Rumsey (1986), who translated the problem of counting TSSCPPs into the counting of a certain class of shifted plane partitions.

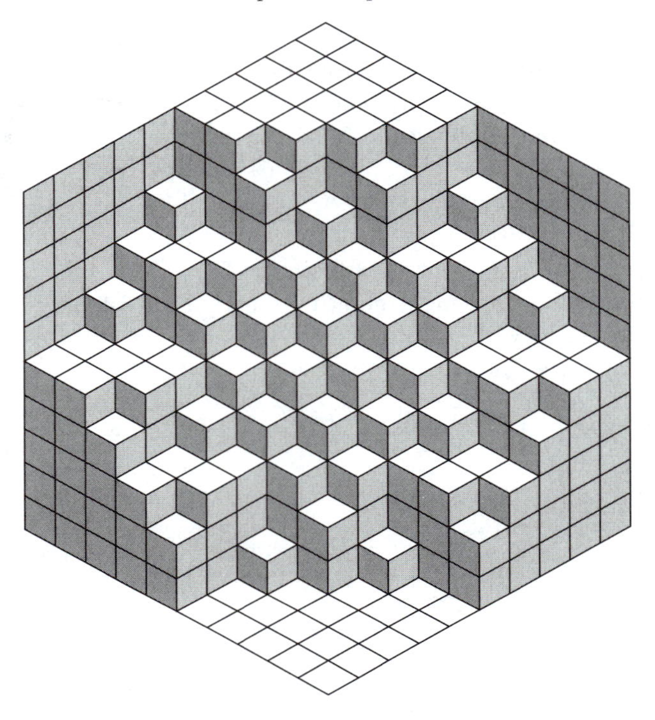

Figure 6.3. A totally symmetric self-complementary plane partition in $\mathcal{B}(10, 10, 10)$.

An example of a TSSCPP in $\mathcal{B}(10, 10, 10)$ is shown in Figure 6.3. As we did with cyclically symmetric plane partitions, we want to decompose our TSSCPPs into shells. Because of the condition that this plane partition must be self-complementary, the outermost shell consists of those lattice points on any of the six exterior faces. For the box $\mathcal{B}(2n, 2n, 2n)$, these are the lattice points for which i, j, or k is equal to 1 or $2n$. The next shell consists of lattice points which are not in the first shell but for which i, j, or k is equal to 2 or $2n - 1$. This continues to the nth shell which consists of those lattice points for which each coordinate is either n or $n + 1$. The entire plane partition will be totally symmetric and self-complementary if and only if each shell is totally symmetric and self-complementary.

We do not have much choice about how to construct a given shell. The outer shell is uniquely determined by the lattice points of the form $(i, j, 2n)$ with $1 \leq i, j \leq n$. Furthermore, these lattice points must form a

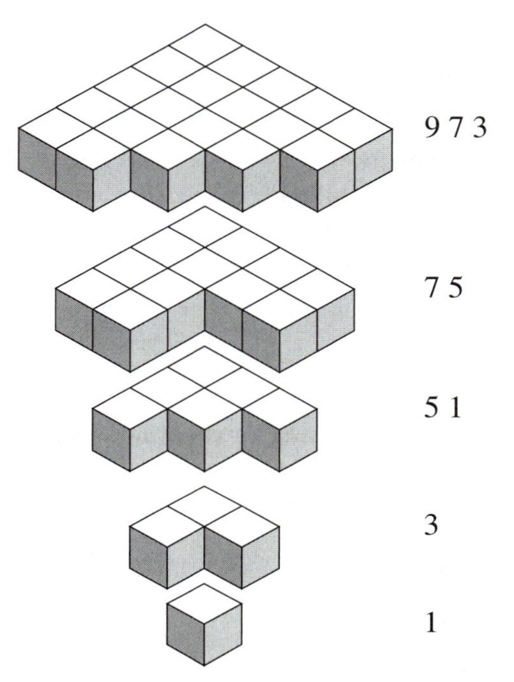

Figure 6.4. The self-conjugate partitions that define the TSSCPP of Figure 6.3.

self-conjugate partition with exactly n parts (and therefore with largest part equal to n). The self-conjugate partitions that define the TSSCPP of Figure 6.3 are given in Figure 6.4. Using the correspondence between self-conjugate partitions and partitions into odd parts (see exercise 2.2.1, page 51), we see that the outermost shell is completely determined by a partition into distinct odd parts of which the largest part must be $2n-1$.

The next shell is determined by a partition into distinct odd parts of which the largest part must be $2n-3$. Since these shells must fit together to form a plane partition, the ith largest part in the second row must be greater than or equal to the $i+1$st largest part in the row above it. As an example, the TSSCPP in Figure 6.3 is described by

$$\begin{array}{ccccc} 9 & 7 & 3 & & \\ & 7 & 5 & & \\ & & 5 & 1 & \\ & & & 3 & \\ & & & & 1. \end{array}$$

We shall call such an array a **TSSCPP array of order** n. If $a_{i,j}$ is the entry in the jth column of the ith row (counted from the top), then an array of positive odd integers is a TSSCPP array of order n if and only if

1. an entry can only occur at position (i, j) if $1 \le i \le j \le n$,
2. for $1 \le i \le n$, $a_{i,i} = 2n + 1 - 2i$,
3. there is strict decrease across rows ($a_{i,j-1} > a_{i,j}$) and weak increase down columns ($a_{i-1,j} \le a_{i,j}$).

There is a one-to-one correspondence between the set of TSSCPP arrays of order n and the set of TSSCPPs that fit inside $\mathcal{B}(2n, 2n, 2n)$.

The corresponding nest of lattice paths

The next step is to translate this array into a nest of lattice paths. This was first done by William F. Doran (1993), who was an undergraduate at the time. The ith row of the TSSCPP array of order n corresponds to a lattice path that starts at $(-2(n-1-i), n-1-i)$, moves to the right or up at each step, and ends when it reaches a lattice point on the line $y = 1 - x$. In constructing this lattice path, we ignore the first entry in the row. If the next entry is $2k - 1$, then we move up until we are at the lattice point that is k units below the line $y = 1 - x$, and then take one step to the right. We repeat this until all entries in the row have been encoded. The lattice path then moves up until it reaches the lattice point on the line $y = 1 - x$. See Figure 6.5.

The nth row of a TSSCPP array of order n always consists of a single 1 which corresponds to the single point lattice path from $(2, -1)$ to $(2, -1)$. As we shall see later, we always want an even number of lattice paths, and so we shall either include or ignore this trivial lattice path depending on whether n is even or odd. The condition of weak increase down columns in the TSSCPP array is equivalent to the condition that the lattice paths do not intersect.

This looks like the kind of counting problem that can be translated into a determinant evaluation by summing over all nests of lattice paths, including those that intersect, and proving that the nests with intersecting paths can be paired into couples with equal weight but opposite sign. This will be our approach, but there is a problem. The methods described in Chapter 3 either involve nests from a given set of n starting points to a set of n ending points or, by adding the identity matrix, from any subset of the starting points to any subset of equal cardinality of the ending points. The problem with the nests that we are now working

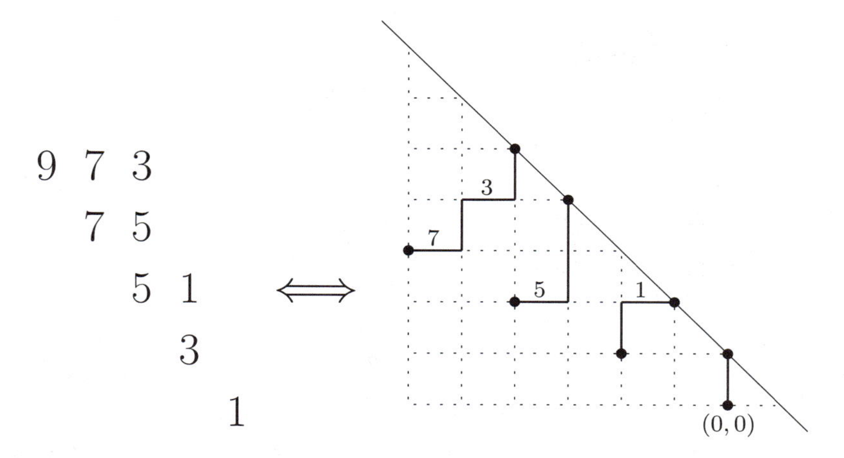

Figure 6.5. A TSSCPP array of order 5 and the corresponding nest of lattice paths.

with is that while we have $n-1$ starting points, the ending points must be chosen from a set of $2n-2$ possible points.

Soichi Okada (1989), working on the problem of counting totally symmetric plane partitions, realized that rather than using determinants which are a sum over all permutations, the right tool in this case is the Pfaffian, a sum over perfect matchings. The use of Pfaffians to count plane partitions is actually much older, going back to work of Basil Gordon (1971). Building on Okada's paper, John Stembridge (1990) showed how Pfaffians could be applied to a wide range of counting problems, including the totally symmetric self-complementary plane partitions.

Pfaffians

Given an even integer n, let $\mathcal{F}_n$ denote the set of all partitions of $\{1, 2, \ldots, n\}$ into pairs of elements. Such a set partition is also called a **perfect matching** or a **1-factor**. As an example, $\mathcal{F}_4$ has three perfect matchings:

$$F_1 = \{(1,2),(3,4)\}; \quad F_2 = \{(1,3),(2,4)\}; \quad \text{and } F_3 = \{(1,4),(2,3)\}.$$

There are fifteen perfect matchings when $n = 6$. Given a perfect matching $F \in \mathcal{F}_n$, we define the **crossing number** of F, $\mathcal{X}(F)$, to be the number of pairs $(a,b),(c,d)$ in F for which

$$a < c < b < d.$$

For example, $\mathcal{X}(\{(1,4),(2,6),(3,5)\}) = 2$ because $(1,4)$ and $(2,6)$ cross and $(1,4)$ and $(3,5)$ cross, but $(2,6)$ and $(3,5)$ do not cross.

Given an ordered collection of $\binom{n}{2}$ quantities, $A = (a_{ij})_{1\leq i<j\leq n}$, the **Pfaffian** of A, $\mathrm{Pf}(A)$, is defined as

$$\mathrm{Pf}(A) = \sum_{F\in\mathcal{F}_n} (-1)^{\mathcal{X}(F)} \prod_{(i,j)\in F} a_{ij}. \tag{6.27}$$

When $n = 4$, the Pfaffian is

$$a_{12}a_{34} - a_{13}a_{24} + a_{14}a_{23}.$$

These sums first appeared in 1815 in a paper by Johann Friedrich Pfaff (1765–1825) that described a method to solve a system of $2n$ first-order differential equations by introducing auxiliary variables and equations. These auxiliary equations involved sums of fractions in which the denominators are given by Pfaffians. Pfaff was Gauss's teacher, and Gauss lived in his house while he was a student. Pfaff coined the term "hypergeometric series" and made many contributions to the study of differential equations.

The Pfaffian is closely related to the determinant of a **skew symmetric matrix** – a matrix for which $a_{ij} = -a_{ji}$ – which appeared in work of Poisson, Lagrange, Laplace, and Monge in the first decade of the nineteenth century. Jacobi (1827, 1845) was the first to recognize that there was a relationship between the Pfaffian and the determinant of a skew symmetric matrix, but it was not until Arthur Cayley (1847a, 1847b) that it was recognized and proven that the determinant of the skew symmetric matrix $(a_{ij})_{i,j=1}^{n}$ is equal to the square of the Pfaffian of $(a_{ij})_{1\leq i<j\leq n}$ (see exercise 6.2.11 for a proof). For example, when $n = 4$ we have

$$\det\begin{pmatrix} 0 & a_{12} & a_{13} & a_{14} \\ -a_{12} & 0 & a_{23} & a_{24} \\ -a_{13} & -a_{23} & 0 & a_{34} \\ -a_{14} & -a_{24} & -a_{34} & 0 \end{pmatrix} = (a_{12}a_{34} - a_{13}a_{24} + a_{14}a_{23})^2.$$

Cayley is also responsible for the terminology "skew symmetric matrix" and "Pfaffian."

There is one other important result about Pfaffians that we need before proceeding. That is that when $a_{ij} = 1$, $1 \leq i < j \leq n$, then the Pfaffian is +1 (see exercise 6.2.9). Following Stembridge, we shall express the number of totally symmetric, self-complementary plane partitions inside $\mathcal{B}(2n, 2n, 2n)$ as a Pfaffian which will then be expressed as the square root of the determinant of the corresponding skew symmetric matrix.

TSSCPP as Pfaffian

Given an integer n, we define $\mathcal{N}_n$ to be the set of all nests of lattice paths for which the ith path starts at $(2-2i, i-1)$, moves to the right or up at each step, and ends when it reaches a lattice point on the line $y = 1 - x$. When n is odd, we take $1 \leq i \leq n-1$. When n is even, we take $0 \leq i \leq n-1$. As we noted above, path $i = 0$ is a trivial path that starts and ends at $(2, -1)$. For every value of n, we have an even number of paths in $\mathcal{N}_n$.

Each nest of lattice paths, $N \in \mathcal{N}_n$, is given a weight, $w(N)$, as follows. For each perfect matching, F, of the initial vertices, we define the inversion number $\mathcal{I}(F, N)$ to be the number of pairs $(i, j) \in F$, $i < j$, for which the end vertex of the ith lattice path is to the left of the end vertex of the jth lattice path. In other words, two lattice paths paired by F are counted in the inversion number if and only if they cross an odd number of times. We define $w(F, N)$ to be 0 if two matched paths end at the same vertex, and $w(F, N) = (-1)^{\mathcal{I}(F,N)}$ otherwise. We define

$$w(N) = \sum_F (-1)^{\mathcal{X}(F)} w(F, N). \tag{6.28}$$

If there are no intersections of lattice paths in N, then $w(F, N) = (-1)^{\mathcal{I}(F,N)} = 1$ for all pairings, and so this sum is the Pfaffian with all variables set equal to 1. In other words, if N has no intersections, then $w(N) = 1$. We shall see that if N has at least one intersection, then for each perfect matching F, we can find a unique corresponding pair (F', N') such that $w(F, N) = -w(F', N')$. It follows that

$$\sum_{N \in \mathcal{N}_n} w(N) = \#\text{ of non-intersecting nests in } \mathcal{N}_n. \tag{6.29}$$

The pairing

Our pairing is similar to the Gessel–Viennot approach described in Section 3.3. If any two paths intersect, then we choose the point of intersection that is closest to one of the starting points, and among those points at the same distance we choose the one with x coordinate closest to 0. There will be exactly two consecutive paths that intersect at this point, say paths i and $i+1$. We create a new nest, N', by interchanging the tails after this point of intersection, and we change the matching from F to F' by exchanging i and $i+1$. We have to prove that $w(N,F) = -w(N',F')$.

If i and $i+1$ are not paired in the matching, then $\mathcal{X}(F) = \mathcal{X}(F') \pm 1$. Whatever was matched with i is now matched with $i+1$, but path $i+1$ now ends where path i used to end. The same observation holds for the number that had been matched with $i+1$, so there is no change in the inversion number.

If i and $i+1$ are paired in the matching, then $\mathcal{X}(F) = \mathcal{X}(F')$. If the intersection point lies on the line $y = 1-x$, then $w(N,F) = -w(N',F') = 0$. If it does not lie on this line, then exchanging tails changes the inversion number by 1.

Sum of weights as Pfaffian

In any nest of lattice paths, the final vertices are chosen from

$$\{(2-r, r-1) \mid 1 \leq r \leq 2n-2,\ n \text{ odd}; 0 \leq r \leq 2n-2,\ n \text{ even}\}.$$

Let $h(i,r)$ be the number of paths that start at the ith vertex, $(2-2i, i-1)$, and end at the rth, $(2-r, r-1)$,

$$h(i,r) = \binom{i}{r-i}.$$

This is zero unless $i \leq r \leq 2i$. If we fix a matching, F, then we can write

$$\sum_{N \in \mathcal{N}_n} w(F,N) = \prod_{(i,j)\in F} \sum_{i \leq r < s \leq 2j} h(i,r)h(j,s) - h(i,s)h(j,r).$$

It follows that

$$\begin{aligned}\sum_{N \in \mathcal{N}_n} w(N) &= \sum_F (-1)^{\mathcal{X}(F)} \sum_{N \in \mathcal{N}_n} w(F,N) \\ &= \sum_F (-1)^{\mathcal{X}(F)} \prod_{(i,j)\in F} H(i,j),\end{aligned}$$

where

$$H(i,j) = \sum_{i \leq r < s \leq 2j} \binom{i}{r-i}\binom{j}{s-j} - \binom{i}{s-i}\binom{j}{r-j}.$$

When we combine this result with equation (6.29), we see that the number of non-intersecting nests in $\mathcal{N}_n$ is the Pfaffian of $(H(i,j))$ where $0 \le i < j \le n-1$ when n is even and $1 \le i < j \le n-1$ when n is odd. We have proven the following result.

Proposition 6.1 *The number of* TSSCPPs *in* $\mathcal{B}(2n, 2n, 2n)$ *is the square root of the determinant of the skew symmetric matrix with entries* $H(i,j)$ *for* 0 or $1 \le i < j < n$, *where the lower limit on* i *is 0 if* n *is even and 1 if* n *is odd.*

The determinant evaluation

The enumeration of totally symmetric self-complementary plane partitions has been reduced to finding the determinant of a certain skew symmetric matrix (and then taking the square root of that determinant). This determinant evaluation was accomplished by George Andrews (1994). While the details are daunting – in Andrews's own words, "the material in this section is inordinately complicated" – the fundamental idea is simplicity itself. To evaluate the determinant of M, we find an upper triangular matrix, R, such that MR is lower triangular. We can scale R so that its determinant is 1, which means that the determinant of M is just the product of the elements on the main diagonal of MR. This particular approach has proven to be extremely fruitful, though its applicability is limited by the ingenuity of the person who is using it. Other mathematicians, especially Christian Krattenthaler and Doron Zeilberger, have developed more systematic – and thus more easily programmable – methods for evaluating such determinants, although their methods require the discovery of additional parameters. A good example of this is described by Krattenthaler (1997), who proves a determinant evaluation that generalizes the evaluation needed for the counting of totally symmetric self-complementary plane partitions.

Andrews did make use of computers in an essential way. One of the key identities required in his proof is

$$\sum_{k=0}^{j} \frac{2^{k+1}(-1)^{j+k}(2j-k)!\,(3k+5)\,(k+j+2)!\,(2k+3)!}{(j-k)!\,(k+1)!\,k!(k+2j+5)!} = 1. \tag{6.30}$$

To get a handle on the left side of this equality, we can write it as a hypergeometric series. It is equal to

$$\frac{(-1)^j 2\,(2j)!\,5\,(j+2)!\,3!}{j!\,(2j+5)!}\;{}_4F_3\left(\begin{matrix}-j,\ 8/3,\ j+3,\ 5/2\\ 5/3,\ -2j,\ 2j+6\end{matrix};-8\right).$$

As Andrews pointed out in his paper, this "is not one of the well-known summable series."†

Fortunately, Doron Zeilberger and Herb Wilf had developed a practical and powerful algorithm for proving just such identities. Known as the **WZ-method**, it builds on earlier work of Sister Mary Celine Fasenmyer (1945) and Bill Gosper (1978). We shall meet it once again at the end of Chapter 7. A description of the WZ-method with illustrations of how it is implemented and pointers to free packages that will run it in either *Mathematica* or *Maple* can be found in the book $A = B$ (Petkovšek, Wilf, Zeilberger 1996).

Exercises

6.2.1 Show that any TSSCPP in $\mathcal{B}(2n,2n,2n)$ must have a stack of height at least $n+1$ at position (n,n).

6.2.2 Write the seven TSSCPPs in $\mathcal{B}(6,6,6)$ using MacMahon's bird's-eye notation (page 93).

6.2.3 Fill in the missing values in the bird's-eye notation for the following TSSCPP in $\mathcal{B}(10,10,10)$:

$$\begin{matrix}
? & ? & ? & ? & ? & ? & ? & ? & ? & ?\\
? & ? & ? & ? & ? & ? & ? & ? & ? & ?\\
? & ? & ? & ? & ? & ? & ? & ? & ? & ?\\
? & ? & ? & ? & ? & ? & ? & ? & ? & ?\\
? & ? & ? & ? & ? & ? & ? & ? & ? & ?\\
? & ? & ? & ? & ? & 3 & 2 & 2 & 1 & ?\\
? & ? & ? & ? & ? & ? & 1 & 1 & 0 & ?\\
? & ? & ? & ? & ? & ? & ? & 0 & ? & ?\\
? & ? & ? & ? & ? & ? & ? & ? & ? & ?\\
? & ? & ? & ? & ? & ? & ? & ? & ? & ?
\end{matrix}$$

6.2.4 Translate the TSSCPP of exercise 6.2.3 into a TSSCPP array and then into a nest of lattice paths.

6.2.5 What are the fifteen perfect matchings of $\{1,2,3,4,5,6\}$?

† In fact, this identity had been discovered by Ira Gessel and Dennis Stanton in 1982, equation 1.8, but it was not widely known.

6.2.6 Prove that $|\mathcal{F}_{2n}| = (2n-1)|\mathcal{F}_{2n-2}|$, and therefore

$$|\mathcal{F}_{2n}| = 1 \cdot 3 \cdot 5 \cdots (2n-1).$$

6.2.7 Prove that if n is odd, then any skew symmetric matrix ($a_{ij} = -a_{ji}$) must have determinant equal to zero.

6.2.8 Given a nest of $2n$ lattice paths, N, prove that $\sum_{F \in \mathcal{F}_{2n}} (-1)^{\mathcal{X}(F)+\mathcal{I}(F,N)}$ cannot be zero.

6.2.9 Define the polynomials $C_n(x) = \sum_{F \in \mathcal{F}_{2n}} x^{\mathcal{X}(F)}$ so that

$$\begin{aligned} C_1(x) &= 1, \\ C_2(x) &= 2 + x, \\ C_3(x) &= 5 + 6x + 3x^2 + x^3. \end{aligned}$$

Prove that $C_n(-1) = 1$ for all $n \geq 1$.

6.2.10 Given a permutation of n letters, σ, the **cycle representation** of σ is an ordered set partition of $\{1, 2, \ldots, n\}$ such that j follows i if and only if $j = \sigma(i)$, and the first element in each ordered set is the image of the last element. For example, the permutation whose sequence representation is given by $\sigma = 246153$ has cycle representation $(1,2,4)(3,6)(5)$. What is the cycle representation of 792341658? What is the sequence representation of $(1,5,2), (3,8), (4,9), (6,7)$?

6.2.11 This exercise and the next constitute Stembridge's proof (1990) that the determinant of a skew symmetric matrix is the square of the Pfaffian. Prove that for any permutation σ, the parity of $\mathcal{I}(\sigma)$ is the same as the parity of the number of cycles of even length.

Let $E_n \subset \mathcal{S}_n$ be the set of permutations on n letters all of whose cycles have even length. Prove that if A is skew symmetric, then

$$\det(A) = \sum_{\sigma \in E_n} (-1)^{\mathcal{I}(\sigma)} \prod_{i=1}^{n} a_{i,\sigma(i)}.$$

6.2.12 We define a bijection between the set of pairs of matchings, $\mathcal{F}_n \times \mathcal{F}_n$, and E_n as follows. Given an ordered pair of matchings, (F, F'), we construct a graph on the vertices 1 through n in which i is connected to j if and only if i is paired with j in F or F'. If i is paired with j in both matchings, then we connect these vertices with two edges. The resulting graph is a disjoint union of cycles of even length. Each cycle will be a cycle of the permutation σ. To determine the direction of each cycle, we take the smallest vertex in the cycle. Its image under σ is the vertex with which it is paired in F. Prove that if (F, F') corresponds to σ, then

$$(-1)^{\mathcal{X}(F)+\mathcal{X}(F')} \prod_{(i,j)\in F} a_{ij} \prod_{(i,j)\in F'} a_{ij} = (-1)^{\mathcal{I}(\sigma)} \prod_{i=1}^{n} a_{i,\sigma(i)}.$$

6.2.13 Prove that for $i \le j$,

$$H(i,j) = \sum_{2i-j<r\le 2j-i} \binom{i+j}{r}. \tag{6.31}$$

Hint: First show that $H(i,j)$ is the number of non-intersecting pairs of lattice paths from the ith and jth vertices to any two vertices on the line $y = 1-x$. Next, show that the total number of pairs of lattice paths from the ith and jth vertices to any two vertices on the line $y = 1-x$ is 2^{i+j}.

Let $b(k,l)$ be the number of pairs of intersecting paths where the first path goes from $(2-2i, 1-i)$ to $(2-l-j, l+j-1)$ and the second path goes from $(2-2j, j-1)$ to $(2-k-j, k+j-1)$. Show that if $k < l$, then

$$b(k,l) = \binom{i}{l+j-i}\binom{j}{k}.$$

Use the tail switch to explain why $b(k,l) = b(l,k)$. It follows that

$$\begin{aligned} H(i,j) &= 2^{i+j} - \sum_{0\le k<l\le 2i-j} \binom{i}{l+j-i}\binom{j}{k} \\ &\quad - \sum_{0\le k\le l\le 2i-j} \binom{i}{l+j-i}\binom{j}{k}. \end{aligned}$$

In the first summation, replace l by $2i-j-l$. In the second summation, replace k by $2i-j-k$. Show that

$$\begin{aligned} H(i,j) &= 2^{i+j} - \sum_{k+l<2i-j} \binom{i}{i-l}\binom{j}{j-k} \\ &\quad - \sum_{k+l\ge 2i-j} \binom{i}{2i-j-l}\binom{j}{2i-j-k} \\ &= 2^{i+j} - \sum_{r>2j-i} \binom{i+j}{r} - \sum_{r\le 2i-j} \binom{i+j}{r}. \end{aligned}$$

6.2.14 The following *Mathematica* code will evaluate the determinant of the skew symmetric matrix defined by $H(i,j)$. Verify up to at least $n = 10$ that this determinant is equal to the square of the $\mathcal{A}_n$.

```
H[i_,j_]:=If[i<=j,Sum[Binomial[i+j,r],
   {r,2i-j+1,2j-i}],-H[j,i]];
Pf[n_]:=Det[Table[Table[H[i,j],
   {i,Mod[n,2],n-1}],{j,Mod[n,2],n-1}]]
```

6.2.15 Verify that

$$\sum_{k=0}^{j} \frac{2^{k+1}(-1)^{j+k}(2j-k)!\,(3k+5)\,(k+j+2)!\,(2k+3)!}{(j-k)!\,(k+1)!\,k!(k+2j+5)!}$$

is equal to

$$\frac{(-1)^j 2\,(2j)!\,5\,(j+2)!\,3!}{j!\,(2j+5)!}\;{}_4F_3\left(\begin{matrix} -j,\ 8/3,\ j+3,\ 5/2 \\ 5/3,\ -2j,\ 2j+6 \end{matrix}\,;-8\right).$$

6.3 Proof of the ASM conjecture

No one had succeeded in finding a one-to-one correspondence between descending plane partitions and alternating sign matrices, but there was reason to believe that it might be possible to find such a correspondence between totally symmetric self-complementary plane partitions (TSSCPPs) and alternating sign matrices. After all, we were now comparing two sets of objects that each had high degrees of symmetry. It was a bit unsettling that in the case of alternating sign matrices this was the symmetry group of the square acting on the entire class of objects, while for TSSCPPs each element was individually invariant under the group of symmetries of a hexagon. But even so, the existence of high degrees of symmetry gave hope that the problem would prove tractable. In anticipation that it might be possible to prove the formula for the number of TSSCPPs in $\mathcal{B}(2n,2n,2n)$, Mills, Robbins, and Rumsey had conjectured a refined correspondence between alternating sign matrices and TSSCPPs.

Given n and a nest of $n-1$ lattice paths that defines a TSSCPP, we consider the $n-1$ sets of lattice squares that lie on or below the line $y = 1 - x$ and to the right of the vertical steps in a given lattice path. We shall call each of these sets a **shape**. Note that a lattice square may contribute to more than one shape, as does the square with center

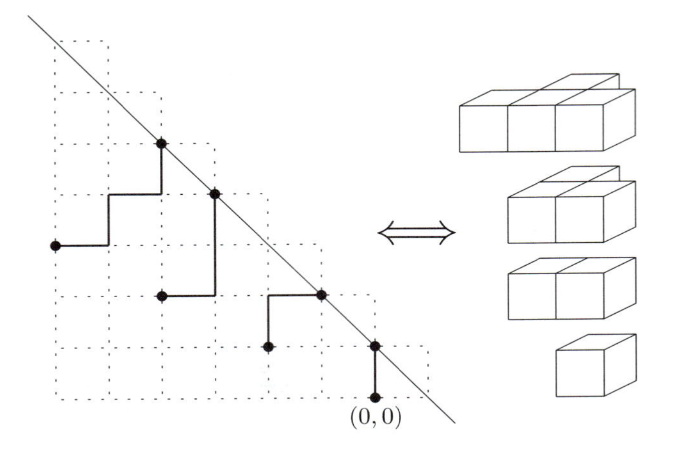

Figure 6.6. A nest of lattice paths and the four corresponding shapes.

at $(-2.5, 3.5)$. The nest of lattice paths in Figure 6.5 is reproduced in Figure 6.6 together with the four shapes that it defines. We use unit cubes to create the shape.

We now create a triangular base, lay the first shape on top of it in the lower right corner, and lay each successive shape on top of the previous (see Fig. 6.7). The fact that our lattice paths were nonintersecting is both necessary and sufficient to guarantee that each shape can be placed above the previous one. We now record this stack of blocks by a triangular arrangement of the integers 1 through n in which the entry in the jth column from the left and the ith row from the top, $j \leq i$, is the number of blocks stacked above that position. We get a triangle with integer entries $c_{ij}, 1 \leq j \leq i \leq n$ for which

$$c_{ij} \leq c_{i+1,j}, \quad c_{ij} \leq c_{i,j+1}, \quad \text{and} \quad c_{ij} \leq j.$$

Following Doron Zeilberger, we shall call such a triangle a **magog triangle** of order n. The process we have gone through is uniquely reversible. For any n, the number of magog triangles of order n is equal to the number of totally symmetric self-complementary plane partitions in $\mathcal{B}(2n, 2n, 2n)$.

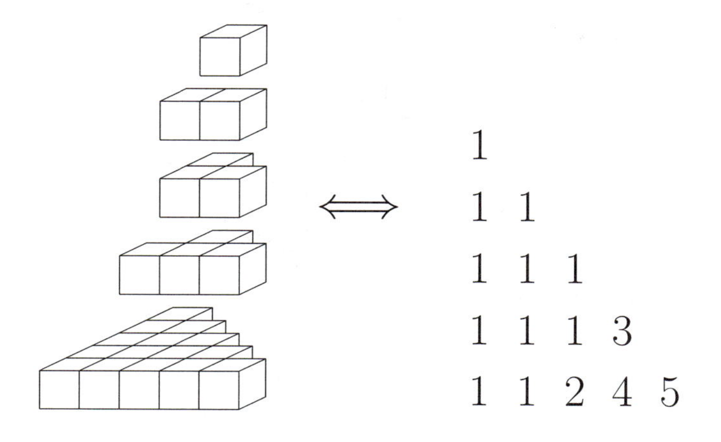

Figure 6.7. The stacked shapes and the corresponding magog triangle.

Gog and magog

We have seen that there is a one-to-one correspondence between $n \times n$ alternating sign matrices and monotone triangles of order n. This suggests that it might be possible to prove the alternating sign matrix conjecture by establishing a one-to-one correspondence between monotone triangles of order n and magog triangles of order n.

While both objects have the integers 1 through n arranged in a triangular arrangement of $n(n+1)/2$ positions, the restrictions are quite different. Mills, Robbins, and Rumsey did observe, however, that the bottom row of a magog triangle consists of a weakly increasing sequences of length n in which the jth entry is less than or equal to j. In a monotone triangle, such as

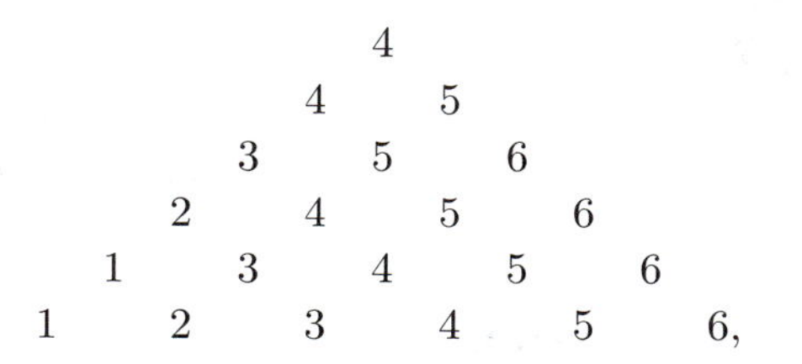

the northwest edge is also a weakly increasing sequence of length n in which the jth entry (as counted from the lower left corner) is less than or equal to j.

What if we take more than just the bottom row of a magog triangle?

We define an (n, k)-**magog trapezoid** to be the bottom k rows of a magog triangle of order n. For example,

$$\begin{array}{cccccc}
1 & 1 & 1 & 1 & & \\
1 & 1 & 1 & 3 & 5 & \\
1 & 1 & 2 & 4 & 5 & 5
\end{array}$$

is a $(6, 3)$-magog trapezoid. We define an (n, k)-**gog trapezoid** to be the first k northwest rows of a monotone triangle of order n. For example,

$$\begin{array}{cccccccc}
 & & & & & 4 & & \\
 & & & & 4 & & 5 & \\
 & & & 3 & & 5 & & 6 \\
 & & 2 & & 4 & & 5 & \\
 & 1 & & 3 & & 4 & & \\
1 & & 2 & & 3 & & &
\end{array}$$

is a $(6, 3)$-gog trapezoid. Mills, Robbins, and Rumsey proved that the number of $(n, 2)$-magog trapezoids is always equal to the number of $(n, 2)$-gog trapezoids and made the following conjecture.

Conjecture 14 *For* $1 \leq k \leq n$*, the number of* (n, k)*-magog trapezoids is equal to the number of* (n, k)*-gog trapezoids.*

Given George Andrews's proof of the formula for the number of totally symmetric self-complementary plane partitions, Conjecture 14 implies the alternating sign matrix conjecture, Conjecture 3. In December of 1992, Doron Zeilberger announced that he had proven Conjecture 14. As it turned out, there were significant gaps in his proof, but by 1995 these had been filled in and checked by one of the largest armies of reviewers that any paper has seen: 88 people and one computer each verified at least one piece of the 71-page proof. The resulting article (Zeilberger 1996a) begins with the list of these 89 referees and concludes with brief descriptions of each of them, a partial "Who's Who" of those exploring some aspect of the alternating sign matrix conjecture in 1995.

Constant term identities

Zeilberger's first step in the proof of Conjecture 14 was to express each of the quantities he wanted to count as the constant term of a Laurent series in k variables, that is to say, a formal power series in k variables and their inverses. The number of (n, k)-magog trapezoids is the constant

Figure 6.8. Doron Zeilberger.

term in

$$F(x_1, \ldots, x_k) = \frac{\prod_{i=1}^k (1-2x_i) \prod_{1 \le i < j \le k} (x_j - x_i)(x_j + \overline{x}_i)}{\prod_{i=1}^k x_i^{n+k-i-1} \overline{x}_i^{n+k+1} \prod_{1 \le i < j \le k} (1 - x_i x_j)},$$

where $\overline{x}_i = 1 - x_i$. The number of (n, k)-gog trapezoids is the constant term in

$$G(x_1, \ldots, x_k) = (-1)^k \frac{\Phi_k(x_1, \ldots, x_k)}{\prod_{i=1}^k x_i^n \overline{x}_i^{n+i+1} \prod_{1 \le i < j \le k} (1 - x_i x_j)(1 - \overline{x}_i x_j)},$$

where $\Phi_k(x_1, \ldots, x_k)$ equals

$$\sum_{\substack{\sigma \in \mathcal{S}_k \\ S \subseteq \{1, \ldots, n\}}} (-1)^{\mathcal{I}(\sigma) + |S|} g_{\sigma,S} \left(\prod_{i=1}^k \overline{x}_i^{k-i} x_i^k \prod_{1 \le i < j \le k} (1 - x_i \overline{x}_j)(1 - \overline{x}_i \overline{x}_j) \right).$$

The operator $g_{\sigma,S}$ acts on a power series in $x_1, \ldots, x_k$ by first replacing each x_i for which $i \in S$ by $\overline{x}_i = 1 - x_i$, and then replacing each x_i by $x_{\sigma(i)}$.

While these expressions may seem dreadful, the reader who goes back to Section 2.4 will see a strong connection to the Weyl denominator formulæ. In making this connection, Zeilberger was able to draw on powerful techniques that had been developed for evaluating constant terms. This thread began with a conjecture made by Freeman Dyson (1962) that for any nonnegative integers $a_1, \ldots, a_n$, the constant term in

$$\prod_{1 \leq i \neq j \leq n} \left(1 - \frac{x_i}{x_j}\right)^{a_j}$$

is equal to the multinomial coefficient

$$\binom{a_1 + a_2 + \cdots + a_n}{a_1,\ a_2,\ \ldots,\ a_n}.$$

I. J. Good gave a particularly simple and transparent proof of this identity (1970) (see exercises 6.3.1–6.3.3).

Ian Macdonald (1982) and others (Andrews 1975, 1980; Askey 1980; Morris 1982) conjectured extensions of this identity that expanded it into the context of the Weyl denominator formula. During the 1980s, most of these conjectures were proven. It was John Stembridge (1988) who realized that the evaluation of such constant terms could be greatly simplified by first summing the series in question over the group of symmetries of the variables. In Zeilberger's proof of the alternating sign matrix conjecture, he first shifts his attention to the residue (the coefficient of $1/x_1 \cdots x_n$) of $F(x_1, \ldots, x_k)/x_1 \cdots x_k$ and $G(x_1, \ldots, x_k)/x_1 \cdots x_k$. He shows that these residues are left unchanged by the operator $g_{\sigma,S}$ and then sums the images of these functions over all pairs (σ, S). *Mirabile dictu*, the two summations turn out to be the same.

The shortcut is discovered

By the start of 1995, all of the bugs had been removed from Zeilberger's proof. It appeared that it should be possible to extend this line of reasoning to find a proof of the refined alternating sign matrix conjecture in which the position of the 1 in the first row is specified, but the details of accomplishing this were daunting. At the same time, Greg Kuperberg discovered a different proof of the alternating sign matrix conjecture.

As a graduate student five years earlier, he had learned from James Propp that physicists were also interested in this problem, though they had not called their structures alternating sign matrices. To the physicists, these were six-vertex models, also known as square ice. Kuperberg learned of the Yang–Baxter equation from Vaughan Jones, which led him to the work of Korepin and Izergin (Korepin, Bogoliubov, and Izergin 1993). In Kuperberg's words (personal communication):

> Actually I approached Vaughan Jones about possible relations between ASM and the Yang–Baxter equation in 1990 or 1991, and in 1991 (I think) Korepin handed me drafts of two chapters of his book. I didn't understand these drafts and their claims seemed overstated to me, so I put them away. I returned to them in 1995 because of Doron's announcements, and for some reason they were much more accessible then. In fact I worked only from the old drafts and not from any published papers.... The potential was clear once Jim Propp had explained the connection with square ice and Vaughan Jones had raved about the Yang–Baxter equation. This all happened when I was in graduate school. It just took a long time for me to accept that it could actually work.

Exercises

6.3.1 Exercises 6.3.1 through 6.3.3 will lead you through I. J. Good's proof of the Dyson conjecture (Good 1970). Prove that for any value of x:

$$\sum_{j=1}^{n} \prod_{\substack{i=1 \\ i \neq j}}^{n} \frac{x - x_i}{x_j - x_i} = 1. \tag{6.32}$$

Hint: The left side of this equality is a polynomial of degree at most $n-1$ in x.

6.3.2 Let

$$F_n(a_1, a_2, \ldots, a_n) = \prod_{1 \leq i \neq j \leq n} \left(1 - \frac{x_j}{x_i}\right)^{a_j}.$$

Use equation (6.32) at $x = 0$ to prove that

$$F_n(a_1, a_2, \ldots, a_n) = \sum_{j=1}^{n} F_n(a_1, a_2, \ldots, a_j - 1, \ldots, a_n).$$

6.3.3 Prove that the constant term of $F_n(a_1, a_2, \ldots, a_n)$ satisfies the same boundary conditions and recursion as the multinomial coefficient, namely,

1. it is symmetric in $a_1, \ldots, a_n$;

2. if $a_n = 0$, then

$$\binom{a_1 + a_2 + \cdots + a_n}{a_1,\ a_2,\ \ldots,\ a_n} = \binom{a_1 + a_2 + \cdots + a_{n-1}}{a_1,\ a_2,\ \ldots,\ a_{n-1}};$$

3. if $a_1, \ldots, a_n$ are all positive, then

$$\binom{a_1 + a_2 + \cdots + a_n}{a_1,\ a_2,\ \ldots,\ a_n} = \sum_{j=1}^{n} \binom{a_1 + a_2 + \cdots + (a_j - 1) + \cdots + a_n}{a_1,\ a_2,\ \ldots,\ a_j - 1,\ \ldots,\ a_n}.$$

6.3.4 There is only one $(n, 1)$-monotone trapezoid (the bottom row of a monotone triangle) and only one leftmost column for a magog triangle. Prove that there are the same number of $(n, 2)$-monotone trapezoids as there are ways of creating the two left-most columns of a magog triangle with parts less than or equal to n.

6.3.5 A general conjecture, which is still unproven, is that the number of (n, k)-monotone trapezoids is equal to the number of ways of selecting the k leftmost columns of a magog triangle with parts less than or equal to n. How far can you verify this conjecture?

6.3.6 More generally, it has been conjectured that the number of trapezoidal pieces k rows high and j columns wide from the southwest corner of a monotone triangle with parts less than or equal to n is equal to the number of rectangular pieces k columns wide and j rows high from the southwest corner of a magog triangle with parts less or equal to n. How far can you verify this conjecture?

7
Square Ice

There are "down-to-earth" physicists and chemists who reject lattice models as being unrealistic. In its most extreme form, their argument is that if a model can be solved exactly, then it must be pathological. I think this is defeatist nonsense.... Admittedly the Ising model has been solved only in one and two dimensions, but two-dimensional systems do exist and can be quite like three-dimensional ones.

– Rodney Baxter (1982)

Statistical mechanics has its origins in the nineteenth-century thermodynamics of Maxwell and his contemporaries, but its modern form was cast in 1902 with the publication by Josiah Willard Gibbs of *Elementary Principles in Statistical Mechanics*. Among the accomplishments of this seminal work is the formula for the free energy of a system of interacting particles,

$$F = -kT \log Z,$$

where k is the Boltzmann constant, T is the temperature, and Z is the **partition function**.

This partition function – which should not be confused with either partitions of integers or set partitions – is a sum, taken over all possible states of the particles, of weights that approach zero as the energy increases. Specifically, we define

$$Z = \sum_{s} e^{-E(s)/kT}, \tag{7.1}$$

where $E(s)$ is the energy of state s, a particular configuration of interacting particles on a lattice. The partition function is central to the study of statistical mechanics. It is used to explain and predict phase transitions and to calculate internal energy as well as the probability that a system will be in a particular state. The explicit calculation of Z is a difficult aspect of statistical mechanics which has been accomplished only for certain very restricted classes of models. Those models for which Z can be calculated exactly are called **exactly solved**.

One important class of models are two-dimensional lattice models in which each position has two possible states (for example, occupied or unoccupied, charged positively or negatively) and in which there is some sort of nearest neighbor interaction (for example, no two adjacent positions can be occupied or the total charge over the four corners of each square must be zero). Even with these restrictions, there are few models that are exactly solved. Square ice, also known as the six-vertex model, is one of them.

7.1 Insights from statistical mechanics

Square ice is a system of water molecules frozen into a square lattice (see Fig. 7.1). Each hydrogen atom lies between two oxygen atoms and may be attached to either one or the other, with the restriction that of the four hydrogen atoms around each oxygen atom, exactly two of them are attached.

If we focus on the oxygen atoms as the vertices of our lattice, each vertex has one of six possible configurations depending on which two of the four surrounding hydrogen atoms are attached to it. This is the origin of the name **six-vertex model**. This model is used both for molecular lattices such as ice and potassium dihydrogen phosphate, KH_2PO_4, and for systems of charged particles such as anti-ferroelectrics. In practice, the six-vertex model is usually drawn as a directed graph on a square lattice as in Figure 7.1 with the arrow pointing toward the oxygen atom to which the particular hydrogen atom is attached. Each vertex has in-degree and out-degree equal to 2.

There is a simple correspondence between $n \times n$ alternating sign matrices and $n \times n$ patches of square ice with the boundary condition that there are hydrogen atoms along the left and right edges and no hydrogen atoms along the top or bottom edges: the horizontal molecules correspond to $+1$, the vertical molecules correspond to -1, and every other molecule corresponds to 0. For each 0 in the alternating sign matrix, we consider the sum of the entries on the same row and to the left and the sum of the entries in the same column and above. If the row sum is 0, then the molecule has a hydrogen atom to the west. If it is 1, there is a hydrogen atom to the east. If the column sum is 0, there is a hydrogen

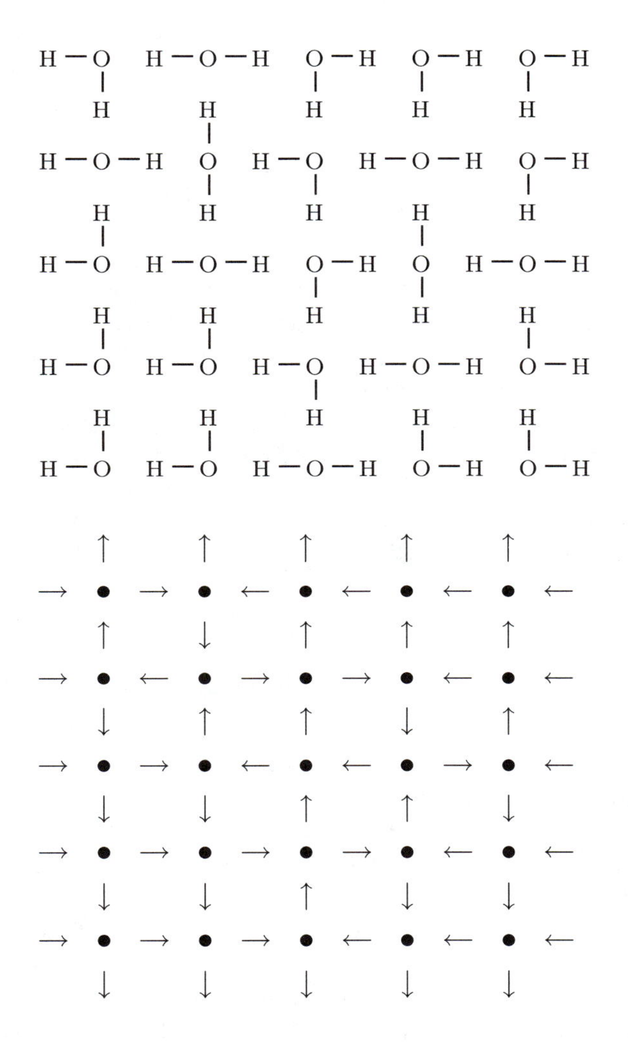

Figure 7.1. Square ice and the corresponding directed graph.

atom to the south. And if it is 1, then there is a hydrogen atom to the north.

Given these boundary conditions, horizontal and vertical molecules must alternate and there must be exactly one more horizontal molecule than vertical in each row or column. Once these two types of molecules have been placed, all of the other molecules are uniquely determined.

The patch of square ice in Figure 7.1 corresponds to the alternating sign matrix

$$\begin{pmatrix} 0 & 1 & 0 & 0 & 0 \\ 1 & -1 & 0 & 1 & 0 \\ 0 & 1 & 0 & -1 & 1 \\ 0 & 0 & 0 & 1 & 0 \\ 0 & 0 & 1 & 0 & 0 \end{pmatrix}.$$

Our boundary restrictions are called **domain wall boundary conditions**. In general, statistical mechanics is concerned with asymptotic behavior as the size of the lattice approaches infinity, and physicists have paid much more attention to the case in which the top and bottom boundaries are identical as are the left and right boundaries. In this case, the sheet of square ice can be wrapped onto a torus and is easier to analyze. Nevertheless, Vladimir Korepin began the serious study of the six-vertex model with domain wall boundary conditions in the early 1980s, showing that the tools developed for toroidal sheets of square ice are also applicable to this case.

Equivalent counting problems

Before continuing with the analysis of square ice, it is worth mentioning that the physicists were aware that their problem was equivalent to other combinatorial problems. In particular, if we take our directed graph, start with the vertices along the left edge, and follow those arrows that go up or to the right, we get a nest of n lattice paths (see Fig. 7.2) in which the ith path goes from $(0, i)$ to $(n - i, n)$. Unlike the nests we studied in Chapter 3, these paths are allowed to touch at corners, although no two paths may actually cross each other or travel along the same edge. The fact that these lattice paths are allowed to intersect means that the methods of Chapter 3 are not directly applicable.

As we shall see, the problem of counting $n \times n$ alternating sign matrices is equivalent to counting the colorings of a square lattice using three colors with the restriction that no two horizontally or vertically adjacent squares may share the same color. We take an $(n+1) \times (n+1)$ arrangement of squares as in Figure 7.3 and impose the following boundary conditions. The square in the upper left-hand corner must have color 1. As we move to the right across the top or down the left edge, the colors cyclically increase by one each time. As we move down the right edge or to the right across the bottom, the colors cyclically decrease by one

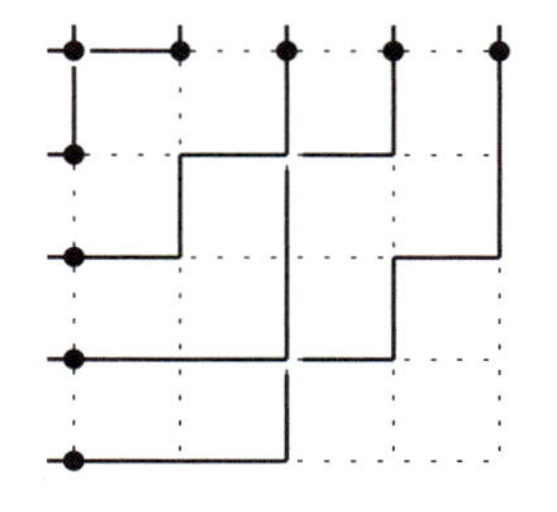

Figure 7.2. The nest of lattice paths corresponding to the directed graph.

$$\begin{array}{ccccccccccc}
1 & \uparrow & 2 & \uparrow & 3 & \uparrow & 1 & \uparrow & 2 & \uparrow & 3 \\
\rightarrow & \bullet & \rightarrow & \bullet & \leftarrow & \bullet & \leftarrow & \bullet & \leftarrow & \bullet & \leftarrow \\
2 & \uparrow & 3 & \downarrow & 2 & \uparrow & 3 & \uparrow & 1 & \uparrow & 2 \\
\rightarrow & \bullet & \leftarrow & \bullet & \rightarrow & \bullet & \rightarrow & \bullet & \leftarrow & \bullet & \leftarrow \\
3 & \downarrow & 2 & \uparrow & 3 & \uparrow & 1 & \downarrow & 3 & \uparrow & 1 \\
\rightarrow & \bullet & \rightarrow & \bullet & \leftarrow & \bullet & \leftarrow & \bullet & \rightarrow & \bullet & \leftarrow \\
1 & \downarrow & 3 & \downarrow & 2 & \uparrow & 3 & \uparrow & 1 & \downarrow & 3 \\
\rightarrow & \bullet & \rightarrow & \bullet & \rightarrow & \bullet & \rightarrow & \bullet & \leftarrow & \bullet & \leftarrow \\
2 & \downarrow & 1 & \downarrow & 3 & \uparrow & 1 & \downarrow & 3 & \downarrow & 2 \\
\rightarrow & \bullet & \rightarrow & \bullet & \rightarrow & \bullet & \leftarrow & \bullet & \leftarrow & \bullet & \leftarrow \\
3 & \downarrow & 2 & \downarrow & 1 & \downarrow & 3 & \downarrow & 2 & \downarrow & 1
\end{array}$$

Figure 7.3. The three-coloring corresponding to the directed graph.

each time. The number of three-colorings of the interior of this lattice is then equal to the number of alternating sign matrices.

It is easiest to describe the correspondence between such a three-coloring and a directed graph. As we move across an edge from one colored square to another, if the arrow points to the left then we increase the color by one; if the arrow points to the right, then we decrease the color by one. The fact that each vertex of the directed graph has in-degree equal to out-degree is equivalent to the statement that as we travel around any vertex, we return to the color from which we started, and thus the coloring is well defined.

The weights

The problem in statistical mechanics is more complicated than just counting the number of possible states. To find the partition function, we must assign a weight to each state. This weight represents the exponential of the energy of that particular state. For the six-vertex model, this weight is the product of weights defined at each vertex. There are four **weight types**, one each for horizontal, vertical, southwest or northeast, and southeast or northwest molecules, respectively:

$$\text{horizontal:} \quad \mathrm{H{-}O{-}H} \quad \text{has weight} \quad z,$$

$$\text{vertical:} \quad \begin{array}{c}\mathrm{H}\\ | \\ \mathrm{O} \\ | \\ \mathrm{H}\end{array} \quad \text{has weight} \quad z^{-1},$$

$$\begin{array}{l}\text{southwest}\\ \text{northeast}\end{array} : \quad \begin{array}{c}\mathrm{H{-}O}\\ \phantom{\mathrm{H{-}}}| \\ \phantom{\mathrm{H{-}}}\mathrm{H}\end{array} \quad \text{or} \quad \begin{array}{c}\mathrm{H}\phantom{\mathrm{{-}H}}\\ |\phantom{\mathrm{{-}H}} \\ \mathrm{O{-}H}\end{array} \quad \text{has weight} \quad \frac{az-(az)^{-1}}{a-a^{-1}},$$

$$\begin{array}{l}\text{southeast}\\ \text{northwest}\end{array} : \quad \begin{array}{c}\mathrm{O{-}H}\\ |\phantom{\mathrm{{-}H}} \\ \mathrm{H}\phantom{\mathrm{{-}H}}\end{array} \quad \text{or} \quad \begin{array}{c}\phantom{\mathrm{H{-}}}\mathrm{H}\\ \phantom{\mathrm{H{-}}}| \\ \mathrm{H{-}O}\end{array} \quad \text{has weight} \quad \frac{z-z^{-1}}{a-a^{-1}}.$$

For convenience, we define

$$[z] = \frac{z-z^{-1}}{a-a^{-1}}.$$

Let $Z_n = Z_n(z,a)$ be the sum of the weights taken over all possible states on an $n \times n$ lattice. As an example,

$$Z_3 = z^3 \left([z]^6 + 2[z]^4 \, [az]^2 + 2[z]^2 \, [az]^4 + [az]^6 + [z]^2 \, [az]^2 \right).$$

Keeping in mind that horizontal molecules correspond to 1s in the alternating sign matrix and vertical molecules correspond to -1s, we see that the total contribution to the weight from the horizontal and vertical molecules is z^n.

We also observe that in the directed graph representation of the six-vertex model, the total number of edges directed up is equal to the total number of edges directed down, and the total number of edges directed to the right is equal to the total number of edges directed to the left. It follows that the number of molecules with hydrogen atoms to the south and west must equal the number of molecules with hydrogen atoms to the north and east. Similarly, the number of molecules with hydrogen atoms to the south and east must equal the number of molecules with hydrogen atoms to the north and west. It follows that the weight of any state must be of the form

$$z^n \, [z]^{2c} \, [az]^{2d},$$

where c and d are integers. I leave it for you (exercise 7.1.7) to prove that $\binom{n}{2} - (c+d)$ is twice the number of -1s in the alternating sign matrix.

If we want to count the total number of alternating sign matrices, we can set

$$z = a = \omega,$$

where ω is a primitive cube root of unity: $\omega = e^{2\pi i/3}$. With these values of z and a, we have that $[z] = [\omega] = 1$ and $[az] = [\omega^2] = -1$, and therefore every state has the same weight:

$$Z_n(\omega, \omega) = \omega^n A_n, \tag{7.2}$$

where A_n is the number of $n \times n$ alternating sign matrices. To evaluate A_n, all we need is an explicit formula for Z_n. Such a formula was found by Anatoli Izergin (1987) who built on Korepin's earlier work. The formula he found is described in Korepin, Bogoliubov, and Izergin's *Quantum Inverse Scattering Method and Correlation Functions* (1993).

In order to derive this formula, we shall introduce more parameters. For the vertex in the ith row and jth column, we replace the z in the corresponding weight by $z_{ij} = x_i/y_j$. We call z_{ij} the **label** of vertex (i, j). We now modify the partition function and define $Z_n(\vec{x}; \vec{y}; a)$ to be the sum over all possible states of the product of the weights of the vertices. The evaluation of the resulting partition function function is given in the following theorem.

Theorem 7.1

$$Z_n(\vec{x}; \vec{y}; a) = \frac{\prod_{i=1}^{n} x_i/y_i \ \prod_{1 \le i,j \le n} [x_i/y_j]\,[a x_i/y_j]}{\prod_{1 \le i < j \le n} [x_i/x_j]\,[y_j/y_i]} \det M, \tag{7.3}$$

where M is the $n \times n$ matrix with entries

$$M_{ij} = \frac{1}{[x_i/y_j]\,[a x_i/y_j]}.$$

Greg Kuperberg saw how to use this formula to calculate A_n. Doron Zeilberger built on Kuperberg's insight and used this formula to prove the refined alternating sign matrix conjecture, Conjectures 1 and 2. Following Zeilberger, we want to set

$$x_1 = \omega t, \quad x_i = \omega \ (2 \le i \le n), \quad y_j = 1 \ (1 \le j \le n).$$

For a given alternating sign matrix, if the 1 in the first row occurs in column r, then the top row consists of a single horizontal molecule of weight ωt, $r-1$ southwest molecules with weight $(\omega^2 t - (\omega^2 t)^{-1})/(\omega - \omega^{-1})$ and $n-r$ southeast molecules with weight $(\omega t - (\omega t)^{-1})/(\omega - \omega^{-1})$. All remaining southeast or northwest molecules will have weight $+1$. The remaining southwest or northeast molecules will have weight -1, and the parity of the number of them will be the same as the parity of $r-1$. The partition function is

$$\begin{aligned} &Z_n(\omega t, \omega, \ldots, \omega; 1, \ldots, 1; \omega) \\ &\quad = \frac{\omega^n t}{(\omega - \omega^{-1})^{n-1}} \\ &\qquad \times \sum_{r=1}^{n} (-1)^{r-1} A_{n,r} \,(t\omega^2 - t^{-1}\omega^{-2})^{r-1}\,(t\omega - t^{-1}\omega^{-1})^{n-r} \qquad (7.4) \end{aligned}$$

where $A_{n,r}$ is the number of $n \times n$ alternating sign matrices with a 1 in the rth column of the first row. Since the polynomials

$$(t\omega^2 - t^{-1}\omega^{-2})^{r-1}\,(t\omega - t^{-1}\omega^{-1})^{n-r}, \quad 1 \le r \le n,$$

are linearly independent, the values of the $A_{n,r}$ are uniquely determined by equation (7.4). If we substitute our conjectured formula for $A_{n,r}$ and it satisfies this equation, then we have proven the conjecture.

Unfortunately, setting $y_j = 1$ and $x_i = \omega$ for all j and all $i > 1$ gives us a highly singular matrix for M and many factors of 0 in the denominator of the product in equation (7.3). We will need to proceed with care. We first set $y_j = q^{(1-j)/2}$, $x_1 = \omega t$, and $x_i = \omega q^{(i-1)/2}$, $i > 1$. We then evaluate the determinant and simplify where possible. Finally, we take the limit $q \to 1$. Our first task, however, is to prove Theorem 7.1.

Exercises

7.1.1 Find the sheet of square ice that corresponds to the alternating sign matrix

$$\begin{pmatrix} 0 & 1 & 0 & 0 \\ 1 & -1 & 1 & 0 \\ 0 & 0 & 0 & 1 \\ 0 & 1 & 0 & 0 \end{pmatrix}.$$

7.1.2 What is the weight (in terms of z and a) of the sheet of square ice in the previous exercise?

7.1.3 Prove that the number of southeast molecules must equal the number of northwest molecules, and that the number of southwest molecules must equal the number of northeast molecules.

7.1.4 Let M be an $n \times n$ alternating sign matrix and let $c_1, c_2, \ldots, c_n$ be the columns of M. Define a new matrix, $S(M)$, whose columns are $(c_1 + c_2 + \cdots c_n, c_2 + \cdots + c_n, \ldots, c_{n-1} + c_n, c_n)$. Show that all entries of $S(M)$ are 0 or 1 and that $S(M)$ has $n(n-1)/2$ 0s and $n(n+1)/2$ 1s.

7.1.5 Let R be an $n \times n$ sheet of square ice. Define the 0-1 matrix $S(R)$ by $s_{ij} = 1$ if and only if the oxygen atom at position ij is attached to the hydrogen atom to its left. Prove that $S(M) = S(R)$.

7.1.6 Prove that in the directed graph that corresponds to a sheet of square ice, the number of edges directed to the right is equal to the number of edges directed to the left.

7.1.7 Prove that if the weight of a lattice is of the form $z^n[z]^{2c}[az]^{2d}$, then the number of -1s in the corresponding alternating sign matrix is $\binom{n}{2} - c - d$.

7.1.8 Prove that the number of southwest or northeast molecules is equal to $2\mathcal{I}(A) - 2N(A)$, where A is the corresponding alternating sign matrix, $\mathcal{I}(A)$ is the inversion number, and $N(A)$ is the number of -1s. Show that the number of northwest or southeast molecules is $n^2 - n - 2\mathcal{I}(A)$.

7.1.9 Prove that $(t\omega^2 - t^{-1}\omega^{-2})^{r-1}(t\omega - t^{-1}\omega^{-1})^{n-r}$, $1 \leq r \leq n$, are linearly independent.

7.1.10 Verify that equation (7.4) is correct for $n = 2$ and 3.

7.1.11 Given an (n, k)-monotone trapezoid, what is the rule to find the corresponding $k - 1$ rows of square ice? Find a recursive definition for $Z_{n,k}(a_1, \ldots, a_k; \omega, \ldots, \omega; 1, \ldots, 1; \omega)$, the sum of all sheets of square ice that correspond to an (n, k)-monotone trapezoid with top row $(a_1, \ldots, a_k)$.

7.1.12 Use the result from the previous exercise to write a *Mathematica* program to compute $Z_n(\omega t, \omega, \ldots, \omega; 1, \ldots, 1; \omega)$.

7.1.13 Consider the weight function, Z, with indeterminate a and z, of a particular sheet of square ice. What happens to this weight function if we reflect the sheet of ice across a vertical line?

7.1.14 What happens to the weight function of a particular sheet of square ice if we rotate the ice by 180°?

7.1.15 The **eight-vertex model** is a directed graph on a square lattice in which each vertex has even in-degree. Thus we have the six types of vertices as before, plus a **sink** (all adjacent arrows are directed toward that vertex) and a **source** (all adjacent arrows are directed away):

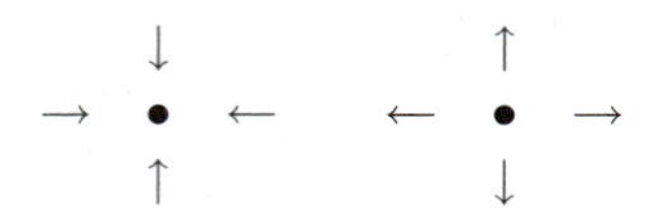

Show that the number of $n \times n$ eight-vertex lattices with the same boundary conditions as before is equal to the number of $n \times n$ matrices of 0s, 1s, -1s, Vs, and Hs with the same alternating conditions as before except that V is read as 1 when going down a column and as -1 across a row, H is read as -1 down a column but as 1 across a row. We shall call these **VH alternating sign matrices**. For example,

$$\begin{pmatrix} 1 & V & 1 \\ H & -1 & H \\ 1 & V & 1 \end{pmatrix}$$

is a VH alternating sign matrix.

7.1.16 Prove that in an eight-vertex model, the number of sources must equal the number of sinks.

7.1.17 Prove that

$$Z_n(z, z^{-2}) = \frac{z^n}{(z^{-1} + z)^{n^2-n}} \sum_{B \in \mathcal{A}_n} (z^{-1} + z)^{2N(B)}. \qquad (7.5)$$

7.2 Baxter's triangle-to-triangle relation

The crucial step in proving Theorem 7.1 is to prove that $Z_n(\vec{x}; \vec{y}; a)$ is a symmetric function in the x_i and is also a symmetric function in the y_j. This is accomplished by using the Yang–Baxter equation for the six-vertex model, what Baxter (in a personal communication) refers to as the triangle-to-triangle relation.

In the language of statistical mechanics, the Yang–Baxter equation for a particular model is the statement that transfer matrices commute. It applies to three classes of lattices: the Ising model (Onsager 1944), the six- and eight-vertex models (Baxter 1971), and the "interactions 'round a face" (IRF) or hard hexagon model (Baxter 1980). In the case of the Ising model, the Yang–Baxter equation is also known as the star-triangle or star-to-triangle relation. The triangle-to-triangle relation applies to both the six- and eight-vertex models. This is the only form of the Yang–Baxter equation that I shall state or prove.

We begin by attaching a new vertex that corresponds to a southwest

Figure 7.4. Rodney Baxter.

molecule, rotated 45° clockwise, along the left edge of our directed graph, between the ith and $i+1$st rows. This introduces a triangle into our lattice.

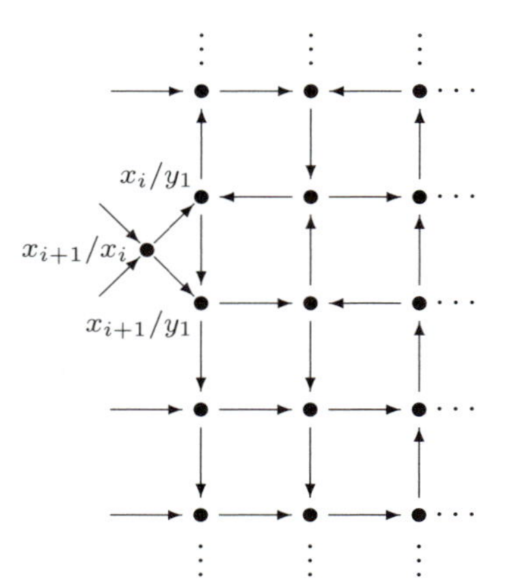

The effect of adding this vertex with label x_{i+1}/x_i is to multiply the weight of the entire lattice by $[ax_{i+1}/x_i]$. We now move the extra vertex across the vertical line while maintaining the directions of the six edges that are incident to the triangle and switching the labels of the top and bottom vertices of the triangle. The weight of a molecule with diagonal arrows is the weight it would have if it were rotated back 45° counterclockwise.

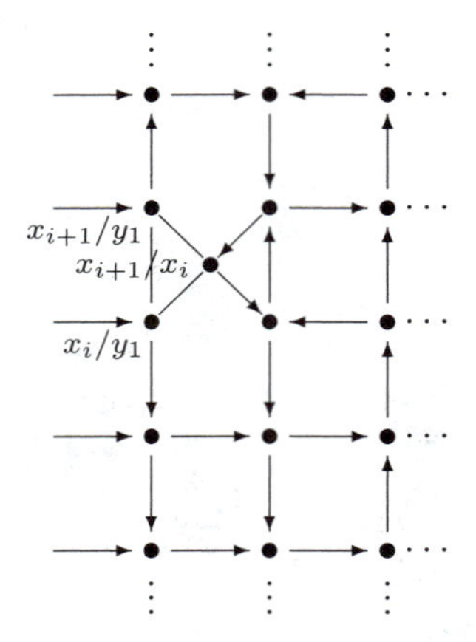

We note that two edges that had been incident to the same vertex of the triangle are now incident to different vertices. There is no change to the direction of the edges incident to any vertex not in the triangle. We also note that in the original triangle, we had no choice for the direction of the edges within the triangle. This is because each vertex must have in-degree and out-degree equal to two. Now that we have moved this extra vertex, we do have a choice. We can go around the triangle either clockwise or counterclockwise:

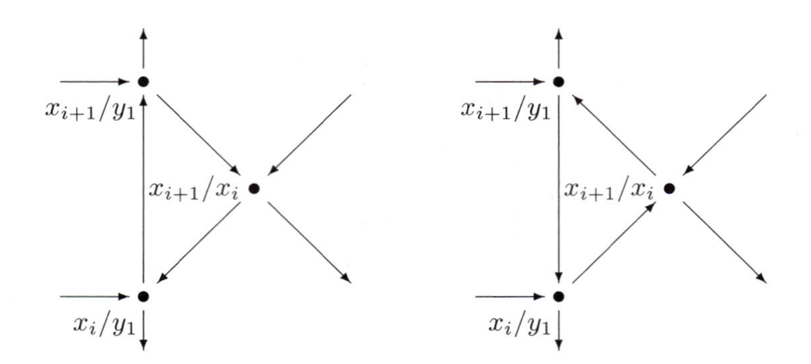

If we sum the two possible weights contributed by the vertices on the

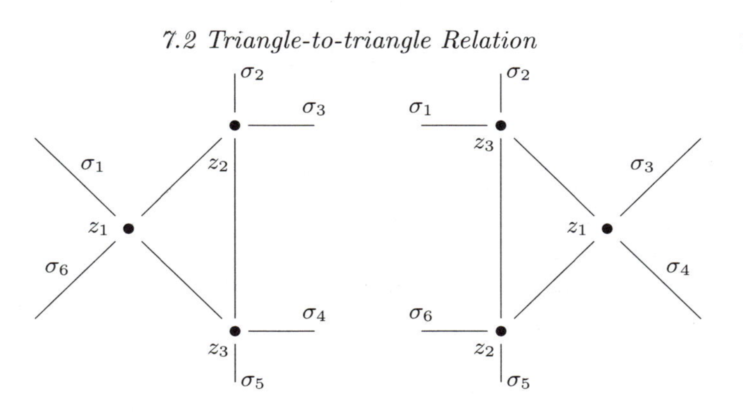

Figure 7.5. The triangle-to-triangle relation.

triangle, we get

$$[x_{i+1}/x_i]\,[ax_{i+1}/y_1]\,\frac{x_i}{y_1} \;+\; \frac{x_i}{x_{i+1}}\,\frac{x_{i+1}}{y_1}\,[x_i/y_1].$$

This happens to be equal to the contribution to the weight from the triangle *before* we moved the extra vertex:

$$[x_{i+1}/x_i]\,[ax_{i+1}/y_1]\,\frac{x_i}{y_1} \;+\; \frac{x_i}{x_{i+1}}\,\frac{x_{i+1}}{y_1}\,[x_i/y_1] = [ax_{i+1}/x_i]\,\frac{x_i}{y_1}\,[x_{i+1}/y_1]. \tag{7.6}$$

This is not a coincidence. The **Yang–Baxter equation for the six-vertex model** or **triangle-to-triangle relation**, given below, states that if we fix any directions for the six edges incident to the triangle – subject only to the condition that three must be directed toward the triangle and three must point away – then the sum of the weights of all possible triangles is the same before and after moving the extra vertex.

Theorem 7.2 (triangle-to-triangle relation) *Given a triangle within a six-vertex lattice and labels z_1, z_2, and $z_3 = z_1z_2$, as in Figure 7.5, we fix the orientations $(\sigma_1, \ldots, \sigma_6)$ of the six edges that are incident to the triangle. Each σ_i is either "in" or "out." If we move the left vertex across to the right, maintain the orientations of the incident edges, and switch the labels of the top and bottom vertices, then the sum of all possible weights under the initial configuration is equal to the sum of all possible weights under the final configuration.*

Proof: Since each of our three vertices has in-degree two and out-degree two, of the six external edges, three must be directed inward and three outward. This gives us twenty possible configurations. Fortunately, many of them are equivalent.

1. We do not change the weight of a molecule if we rotate it by 180°. This means that if we have proven the theorem for one set of orientations and then we rotate the entire picture given in Figure 7.5 by 180°, the theorem must still be correct for this set of orientations. This means that we can replace each σ_i by σ_{i+3}, where addition is taken modulo 6.
2. If we reflect our triangle across the perpendicular bisector that passes through z_3, then we interchange the labels z_1 and z_2, molecules that are not horizontal or vertical do not change their weight type, vertical molecules become horizontal, and horizontal molecules become vertical. This has the same effect as relabeling:

$$z_1 \longrightarrow z_2^{-1}, \quad z_2 \longrightarrow z_1^{-1}, \quad z_3 \longrightarrow z_3^{-1}, \quad a \longrightarrow a^{-1}.$$

   This does not affect the validity of the theorem. This means that we can replace each σ_i by σ_{3-i}, where again addition is taken modulo 6.
3. If we reverse all of the orientations, then the labels remain the same, molecules that are not horizontal or vertical molecules do not change their weight, vertical molecules become horizontal, and horizontal molecules become vertical. This has the same effect as relabeling:

$$z_1 \longrightarrow z_1^{-1}, \quad z_2 \longrightarrow z_2^{-1}, \quad z_3 \longrightarrow z_3^{-1}, \quad a \longrightarrow a^{-1}.$$

   This does not affect the validity of the theorem, and therefore we can replace each σ_i by its opposite.

Together, these three symmetries separate our twenty configurations into five classes of equivalent identities. I leave it for you to identify these equivalence classes (see exercise 7.2.2). We shall prove the triangle-to-triangle relation by verifying it for one representative of each of these classes.

1. The in-edges are σ_1, σ_3, and σ_6. This is the example that has already been done in which the triangle-to-triangle relation reduces to

$$[z_1]\, z_2\, [az_3] + z_1^{-1}\, [z_2]\, z_3 = [az_1]\, z_2\, [z_3]. \tag{7.7}$$

2. The in-edges are σ_1, σ_2, and σ_3. In this case, the triangle-to-triangle relation states that

$$[z_1]\,[az_2]\,[z_3] = [z_1]\,[az_2]\,[z_3]. \tag{7.8}$$

3. The in-edges are σ_1, σ_3, and σ_4. In this case, the triangle-to-triangle relation states that

$$z_1\,z_2\,[az_3] + [z_1]\,[z_2]\,z_3 = [az_1]\,[az_2]\,z_3. \tag{7.9}$$

4. The in-edges are σ_1, σ_3, and σ_5. In this case, the triangle-to-triangle relation states that

$$z_1\,z_2\,z_3^{-1} + [z_1]\,[z_2]\,[az_3] = z_1^{-1}\,z_2^{-1}\,z_3 + [z_1]\,[z_2]\,[az_3]. \tag{7.10}$$

5. The in-edges are σ_2, σ_3, and σ_4. In this case, the triangle-to-triangle relation states that

$$[az_1]\,[az_2]\,[az_3] = [az_1]\,[az_2]\,[az_3]. \tag{7.11}$$

Since $z_3 = z_1 z_2$, the only equalities that require some calculation are equations (7.7) and (7.9). Equation (7.7) is equivalent to (7.6) which you are asked to prove in exercise 7.2.1. In exercise 7.2.4, I ask you to prove equation (7.9).

Q.E.D.

We now return to our original lattice to which we added the extra vertex along the left edge. We have seen that we can move this vertex across the vertical edge to its right and interchange the labels above and below without changing the partition function. The sum of the weights of all possible configurations with this extra vertex on the left edge is equal to the sum of the weights of all possible configurations with the extra vertex moved into the space to its right.

By the same argument, we can move our vertex across the edge that now sits to its right and exchange the labels x_i/y_2 and x_{i+1}/y_2 without changing the partition function. We continue until the extra vertex comes out the right side. As it has moved across, it has interchanged the labels on the vertices in the ith and $i+1$st rows. If we remove the extra vertex from the right-hand edge, we divide the weight of the entire lattice by $[ax_{i+1}/x_i]$, precisely the amount by which we multiplied the partition function when we introduced the extra vertex. Therefore, we have interchanged the labels on rows i and $i+1$ without changing the partition function. A similar argument works for interchanging the labels on two adjacent columns. Since any permutation can be built out of adjacent transpositions, we have proven the following result.

Lemma 7.3 *The partition function $Z_n(\vec{x}, \vec{y}, a)$ is a symmetric function in the x_i, and it is a symmetric function in the y_j.*

We are almost done with the proof of Theorem 7.1. We shall prove it by induction on n. It is easy to verify that it is correct when $n = 1$ or 2. Let $F_n(\vec{x}; \vec{y}; a)$ be the rational product:

$$\frac{\prod_{i=1}^{n} x_i/y_i \prod_{1\le i,j\le n}[x_i/y_j]\,[ax_i/y_j]}{\prod_{1\le i<j\le n}[x_i/x_j]\,[y_j/y_i]} \det\left(\frac{1}{[x_i/y_j]\,[ax_i/y_j]}\right).$$

We need to prove that $F_n = Z_n$. The first thing we do is to get rid of the bracket notation in F_n:

$$\begin{aligned} F_n(\vec{x}; \vec{y}; a) &= \frac{\prod_{i=1}^{n} x_i^{2-n} y_i^{-n}}{(a^2-1)^{n(n-1)}} \frac{\prod_{1\le i,j\le n}(x_i^2 - y_j^2)\,(a^2x_i^2 - y_j^2)}{\prod_{1\le i<j\le n}(x_i^2 - x_j^2)\,(y_j^2 - y_i^2)} \\ &\quad \times \det\left(\frac{1}{(x_i^2 - y_j^2)\,(a^2x_i^2 - y_j^2)}\right). \end{aligned}$$

The product,

$$\left(\prod_{1\le i,j\le n}(x_i^2 - y_j^2)\,(a^2x_i^2 - y_j^2)\right) \det\left(\frac{1}{(x_i^2 - y_j^2)\,(a^2x_i^2 - y_j^2)}\right),$$

is an alternating polynomial in the x_i^2, and it is also an alternating polynomial in the y_j^2. By Proposition 2.8 on page 65,

$$\frac{\prod_{1\le i,j\le n}(x_i^2 - y_j^2)\,(a^2x_i^2 - y_j^2)}{\prod_{1\le i<j\le n}(x_i^2 - x_j^2)\,(y_j^2 - y_i^2)} \det\left(\frac{1}{(x_i^2 - y_j^2)\,(a^2x_i^2 - y_j^2)}\right)$$

is a symmetric polynomial in the x_i and in the y_j. It follows that

$$x_1^{n-2} F_n(\vec{x}; \vec{y}; a)$$

is a polynomial of degree at most $n-1$ in x_1^2, and it is a symmetric function in the y_j.

We now turn our attention to the partition function $Z_n(\vec{x}; \vec{y}; a)$. The first row of our lattice must contain a vertex of weight x_1/y_j for some j, and all other vertices in the first row have weight $[x_1/y_j] = (a/x_1y_j)(x_1^2 - y_j^2)/(a^2-1)$ or $[ax_1/y_j] = (1/x_1y_j)(a^2x_1^2 - y_j^2)/(a^2-1)$. It follows that

$$x_1^{n-2} Z_n(\vec{x}; \vec{y}; a)$$

is a polynomial of degree $n-1$ in x_1^2. By Lemma 7.3, it is a symmetric function in the y_j. Specifically, we now see that

$$P_1(x_1) = x_1^{n-2} Z_n(\vec{x};\vec{y};a) \qquad \text{and} \qquad P_2(x_1) = x_1^{n-2} F_n(\vec{x};\vec{y};a)$$

are both polynomials of degree at most $n-1$ in x_1^2. If we can show that they agree for n distinct values of x_1^2, then they must be equal.

If we set $x_1 = y_1/a$, then the weight of the vertex in the first row and first column of the lattice will be zero unless it is a horizontal molecule of weight $x_1/y_1 = a^{-1}$. In this case, the jth vertex in the first row, $j \geq 2$, will have to be a southeast molecule of weight $[x_1/y_j] = [y_1/ay_j]$, and the ith vertex in the first column, $i \geq 2$, will have to be a northwest molecule of weight $[x_i/y_1]$. The sum over all possible weights of the remainder of the lattice is simply the partition function

$$Z_{n-1}(x_2,\ldots,x_n;y_2,\ldots,y_n;a).$$

We have shown that

$$\begin{aligned} P_1(y_1/a) &= y_1^{n-2} a^{1-n} \left(\prod_{i\neq 1} [y_1/ay_i]\,[x_i/y_1] \right) \\ &\qquad \times Z_{n-1}(x_2,\ldots,x_n;y_2,\ldots,y_n;a). \end{aligned}$$

Since P_1 is symmetric in the y_k, we have that

$$\begin{aligned} P_1(y_k/a) &= y_k^{n-2} a^{1-n} \left(\prod_{i\neq k} [y_k/ay_i] \prod_{i\neq 1} [x_i/y_k] \right) \\ &\qquad \times Z_{n-1}(x_2,\ldots,x_n;\vec{y}\backslash y_k;a), \end{aligned}$$

where $\vec{y}\backslash y_k$ denotes the $n-1$-dimensional vector $\vec{y}$ with y_k removed. On the other hand,

$$\begin{aligned} P_2(y_k/a) &= y_k^{n-2} a^{1-n} \frac{\prod_{j\neq k}[y_k/ay_j]\,[y_k/y_j] \prod_{i\neq 1}[x_i/y_k]\,[ax_i/y_k]}{(-1)^{n-k} \prod_{j\neq 1}[y_k/ax_j] \prod_{j\neq k}[y_k/y_j]} \\ &\qquad \times (-1)^{k+1} F_{n-1}(x_2,\ldots,x_n;\vec{y}\backslash y_k;a) \\ &= y_k^{n-2} a^{1-n} \left(\prod_{i\neq k} [y_k/ay_i] \prod_{i\neq 1} [x_i/y_k] \right) \\ &\qquad \times F_{n-1}(x_2,\ldots,x_n;\vec{y}\backslash y_k;a). \end{aligned}$$

Since our induction hypothesis is that $F_{n-1} = Z_{n-1}$, we see that

$$P_1(y_k/a) = P_2(y_k/a), \qquad 1 \leq k \leq n.$$

We have our n values of x_1^2 at which these polynomials agree, and so they must be identically equal. This concludes the proof of Theorem 7.1.

Q.E.D.

Greg Kuperberg (1996b) saw how to get from Theorem 7.1 to the proof of Conjecture 3. You can follow his steps in exercises 7.2.6 through 7.2.8. But we are now heading for a higher prize, the proof of Conjecture 1.

The partition function specialized

We are ready to specialize our partition function so that we can verify our conjectured formula for $A_{n,r}$. Recall from Section 7.1 that if we set all $x_i = \omega = e^{2\pi i/3}$, all $y_j = 1$, and $a = \omega$, then the weight of each sheet of square ice is ω^n. We restrict our attention to those sheets that have a horizontal molecule in the rth column of the first row. The first $r-1$ molecules in this row have weight -1, the rth molecule has weight ω, and the last $n-r$ molecules have weight 1. It follows that the product of the weights of all vertices below the first row is $(-1)^{r-1}\omega^{n-1}$.

We now introduce an extra parameter so that the determinant in Theorem 7.1 is not zero. We set

$$\begin{aligned} x_1 &= \omega t, \\ x_i &= \omega q^{(i-1)/2}, \quad 2 \le i \le n, \\ y_j &= q^{(1-j)/2}, \quad 1 \le j \le n, \text{ and} \\ a &= \omega. \end{aligned}$$

Let $A_{n,r}(q)$ be $(-1)^{r-1}\omega^{1-n}$ times the sum, taken over all patches of square ice with a horizontal molecule in the rth column of the first row, of the product of the weights of all molecules in all rows below the first. When q approaches 1, we have that

$$\lim_{q \to 1} A_{n,r}(q) = A_{n,r}.$$

On the one hand, we can write our partition function as

$$\begin{aligned} Z_n &= \sum_{r=1}^{n} \omega t q^{(r-1)/2} [\omega^2 t]\, [\omega^2 t q^{1/2}] \cdots [\omega^2 t q^{(r-2)/2}] \\ &\quad \times [\omega t q^{r/2}] \cdots [\omega t q^{(n-1)/2}]\, \omega^{n-1} (-1)^{r-1} A_{n,r}(q). \qquad (7.12) \end{aligned}$$

On the other hand, from equation (7.3), we have that

$$\begin{aligned} Z_n &= \omega^n t q^{n(n-1)/2} \frac{\prod_{j=1}^n [\omega t q^{(j-1)/2}]\,[\omega^2 t q^{(j-1)/2}]}{\prod_{j=2}^n [t q^{(1-j)/2}]\,[q^{(1-j)/2}]} \\ &\quad \times \frac{\prod_{i=2}^n \prod_{j=1}^n [\omega q^{(i+j-2)/2}]\,[\omega^2 q^{(i+j-2)/2}]}{\prod_{2\le i<j\le n} [q^{(i-j)/2}]^2} \det M, \qquad (7.13) \end{aligned}$$

where

$$\begin{aligned} M_{1,j} &= \frac{1}{[\omega t q^{(j-1)/2}]\,[\omega^2 t q^{(j-1)/2}]}, \quad 1 \le j \le n, \\ M_{i,j} &= \frac{1}{[\omega q^{(i+j-2)/2}]\,[\omega^2 q^{(i+j-2)/2}]}, \quad 2 \le i \le n,\ 1 \le j \le n. \end{aligned}$$

We now set these two expression for Z_n equal to each other and divide through by $\omega^n t$:

$$\begin{aligned} &\sum_{r=1}^n (-1)^{r-1} q^{(r-1)/2} [\omega^2 t]\,[\omega^2 t q^{1/2}] \cdots [\omega^2 t q^{(r-2)/2}] \\ &\quad \times [\omega t q^{r/2}] \cdots [\omega t q^{(n-1)/2}]\, A_{n,r}(q) \\ &\qquad = q^{n(n-1)/2} \frac{\prod_{j=1}^n [\omega t q^{(j-1)/2}]\,[\omega^2 t q^{(j-1)/2}]}{\prod_{j=2}^n [t q^{(1-j)/2}]\,[q^{(1-j)/2}]} \\ &\qquad\quad \times \frac{\prod_{i=2}^n \prod_{j=1}^n [\omega q^{(i+j-2)/2}]\,[\omega^2 q^{(i+j-2)/2}]}{\prod_{2\le i<j\le n} [q^{(i-j)/2}]^2} \det M. \qquad (7.14) \end{aligned}$$

We shall get rid of the bracket notation. I leave it for you (exercise 7.2.11) to verify that

$$[\omega t q^i] = \omega^2 t^{-1} q^{-i} \frac{1 - \omega^2 t^2 q^{2i}}{\omega^2 - \omega}, \qquad (7.15)$$

$$[\omega t q^i]\,[\omega^2 t q^i] = \frac{t^{-2} q^{-2i}}{-3} \frac{1 - t^6 q^{6i}}{1 - t^2 q^{2i}}. \qquad (7.16)$$

You should then check (exercise 7.2.12) that equation (7.14) is equivalent to

$$q^{-n(n-1)/4}\omega^{2n-1}\sum_{r=1}^{n}(-1)^{r-1}q^{r-1}\omega^{2r}(1-\omega t^2)(1-\omega t^2 q)\cdots$$

$$\times\,(1-\omega t^2 q^{r-2})(1-\omega^2 t^2 q^r)\cdots(1-\omega^2 t^2 q^{n-1})\,A_{n,r}(q)$$

$$= \frac{q^{-n(n-1)(5n-4)/6}}{(-3)^{(n-1)^2/2}}\,\frac{\prod_{j=1}^{n}(1-t^6q^{3j-3})/(1-t^2q^{j-1})}{\prod_{j=2}^{n}(1-t^2q^{1-j})(1-q^{1-j})}$$

$$\times\,\frac{\prod_{i=2}^{n}\prod_{j=1}^{n}(1-q^{3i+3j-6})/(1-q^{i+j-2})}{\prod_{1\le i<j\le n-1}(1-q^{j-i})^2}\;\det N(n;t), \quad (7.17)$$

where $\sqrt{-3}=\omega^2-\omega$ and $N(n;t)$ is the $n\times n$ matrix whose first row is given by

$$\frac{1-t^2q^{j-1}}{1-t^6q^{3j-3}}, \quad 1\le j\le n,$$

and whose ith row, $2\le i\le n$, is

$$\frac{1-q^{i+j-2}}{1-q^{3i+3j-6}}, \quad 1\le j\le n.$$

If we could evaluate $N(n;t)$ and then take the limit of each side as q approaches 1, then we could prove the refined alternating sign matrix conjecture just by verifying that our conjectured formula for $A_{n,r}$ satisfies this equation. It will prove to be slightly easier to divide each side by $A_{n,n}$ before taking the limit.

Exercises

7.2.1 Prove equation (7.6):

$$[x_{i+1}/x_i]\,[ax_{i+1}/y_1]\,\frac{x_i}{y_1} + \frac{x_i}{x_{i+1}}\,\frac{x_{i+1}}{y_1}\,[x_i/y_1]$$
$$= [ax_{i+1}/x_i]\,\frac{x_i}{y_1}\,[x_{i+1}/y_1].$$

7.2.2 Classify the twenty different configurations of in and out edges around a triangle into five equivalence classes using the equivalences described on page 236. For example, let $\{i,j,k\}$ be the choice of orientations in which σ_i, σ_j, and σ_k are directed in and the other three edges are directed out. The equivalence class that contains $\{1,2,3\}$ also has $\{4,5,6\}$, $\{6,1,2\}$, and $\{3,4,5\}$.

7.2.3 Verify that equation (7.3) is correct for $n=1$ and $n=2$.

7.2.4 Prove equation (7.9).

7.2.5 Verify that $P_2(y_j)$ is equal to

$$y_j^{n-1}\left(\prod_{i\neq j}[ay_j/y_i]\prod_{i\neq 1}[x_i/y_j]\right)F_{n-1}(x_2,\ldots,x_n;\vec{y}\backslash y_j;a).$$

7.2.6 Use the fact (exercise 7.1.17) that

$$\sum_{B\in\mathcal{A}_n}(z^{-1}+z)^{2N(B)}=\frac{(z^{-1}+z)^{n^2-n}}{z^n}Z_n(z,z^{-2})$$

to prove that if we set $x_i = zq^{(i-1)/2}$ and $y_i = q^{-i/2}$ for $1\le i\le n$, then

$$\begin{aligned}(z-z^{-1})^{n^2-n}&\sum_{B\in\mathcal{A}_n}(z^{-1}+z)^{2N(B)}\\ &= \lim_{q\to 1}\frac{\prod_{1\le i,j\le n}(1-z^2q^{i+j-1})(1-z^{-2}q^{i+j-1})}{\prod_{1\le i<j\le n}(1-q^{j-i})^2}\\ &\qquad\times\det\left(\frac{1}{(1-z^2q^{i+j-1})(1-z^{-2}q^{i+j-1})}\right).\end{aligned}\tag{7.18}$$

7.2.7 Show that equation (7.18) with $z=\omega=e^{2\pi i/3}$ yields the following formula for A_n, the number of $n\times n$ alternating sign matrices:

$$\begin{aligned}(-3)^{-\binom{n}{2}}\lim_{q\to 1}&\prod_{1\le i<j\le n}(1-q^{j-i})^{-2}\prod_{1\le i,j\le n}\frac{1-q^{3(i+j-1)}}{1-q^{i+j-1}}\\ &\times\det\left(\frac{1-q^{i+j-1}}{1-q^{3(i+j-1)}}\right).\end{aligned}\tag{7.19}$$

7.2.8 Use the formula given in equation (2.33) on page 72 to evaluate the determinant in equation (7.19), and finish the proof of the alternating sign matrix conjecture, Conjecture 3.

7.2.9 Show that equation (7.18) with $z=e^{\pi i/4}$ yields the following formula:

$$\begin{aligned}(-2)^{\binom{n}{2}}\sum_{B\in\mathcal{A}_n}2^{N(B)} &= \lim_{q\to 1}\prod_{1\le i<j\le n}(1-q^{j-i})^{-2}\\ &\quad\times\prod_{1\le i,j\le n}\frac{1-q^{2(i+j-1)}}{1-q^{i+j-1}}\det\left(\frac{1-q^{i+j-1}}{1-q^{2(i+j-1)}}\right).\end{aligned}$$

7.2.10 Show that equation (7.18) with $z = e^{\pi i/6}$ yields the following formula:

$$(-1)^{\binom{n}{2}} \sum_{B\in\mathcal{A}_n} 3^{N(B)} = \lim_{q\to 1} \prod_{1\le i<j\le n} (1-q^{j-i})^{-2} \times \prod_{1\le i,j\le n} \frac{1+q^{3(i+j-1)}}{1+q^{i+j-1}} \det\left(\frac{1+q^{i+j-1}}{1+q^{3(i+j-1)}}\right).$$

7.2.11 Verify equations (7.15) and (7.16).

7.2.12 Verify that the left side of equation (7.14) is equal to

$$\frac{q^{-n(n-1)/4}}{(\omega^2-\omega)^{n-1}} \sum_{r=1}^{n} (-1)^{r-1} q^{r-1} t^{1-n} \omega^{r-1} \omega^{2(n-r)} \times (1-\omega t^2)(1-\omega t^2 q)\cdots(1-\omega t^2 q^{r-2}) \times (1-\omega^2 t^2 q^r)\cdots(1-\omega^2 t^2 q^{n-1})\, A_{n,r}(q).$$

Verify that the product in front of the determinant on the right side of equation (7.14) is equal to

$$\frac{t^{-1-n} q^{-n(n-1)/2+n(n-1)(n-2)/6-n(n-1)(2n-1)/2}}{(-3)^{n(n+1)/2}} \times \frac{\prod_{j=1}^{n}(1-t^6 q^{3j-3})/(1-t^2 q^{j-1})}{\prod_{j=2}^{n}(1-t^2 q^{1-j})(1-q^{1-j})} \times \frac{\prod_{i=2}^{n}\prod_{j=1}^{n}(1-q^{3i+3j-6})/(1-q^{i+j-2})}{\prod_{1\le i<j\le n-1}(1-q^{j-i})^2}$$

Verify that the determinant of M is equal to

$$(-3)^n t^2 q^{n(n-1)} \det N(n;t).$$

Now put these together to complete the verification of equation (7.17).

7.2.13 Using exercise 7.1.8, show that Theorem 7.1 is equivalent to

$$\det\left(\frac{1}{(x_i+y_j)(ax_i+y_j)}\right) \frac{\prod_{i,j=1}^{n}(x_i+y_j)(ax_i+y_j)}{\prod_{1\le i<j\le n}(x_i-x_j)(y_i-y_j)} = \sum_{A\in\mathcal{A}_n} (-1)^{N(A)}(1-a)^{2N(A)} a^{\binom{n}{2}-\mathcal{I}(A)} \times \prod_{i=1}^{n} x_i^{N_i(A)} y_i^{N^i(A)} \prod_{\substack{1\le i,j\le n\\ a_{ij}=0}} (\alpha_{ij}x_i+y_j), \qquad (7.20)$$

where $N_i(A)$ (respectively, $N^i(A)$) is the number of -1s in row (respectively, column) i and

$$\alpha_{ij} = \begin{cases} a, & \text{if } \sum_{k<j} a_{ik} = \sum_{k<i} a_{kj}, \\ 1, & \text{otherwise.} \end{cases}$$

7.2.14 Show that equation (7.20) with $a = 1$ is equivalent to

$$\det\left(\frac{1}{(x_i+y_j)^2}\right) \frac{\prod_{i,j=1}^{n}(x_i+y_j)}{\prod_{1\le i<j\le n}(x_i-x_j)(y_i-y_j)} = \sum_{\sigma\in\mathcal{S}_n} \prod_{i=1}^{n} \frac{1}{x_i+y_{\sigma(i)}}. \tag{7.21}$$

7.2.15 In equation (7.20), set $a = -1$, $x_i = x$, and $y_i = y$ for all i. Let $\lambda = -(x+y)^2/(x-y)^2$. Show that this equation becomes

$$\sum_{A\in\mathcal{A}_n} \lambda^{\mathcal{I}(A)}(1+\lambda^{-1})^{N(A)} = (1+\lambda)^{\binom{n}{2}}.$$

7.3 Proof of the refined ASM conjecture

Our quest nears its end. In this section we shall use equation (7.17) to prove that

$$\frac{A_{n,r}}{A_{n,n}} = \binom{n+r-2}{n-1} \frac{(2n-r-1)!\,(n-1)!}{(n-r)!\,(2n-2)!}. \tag{7.22}$$

This is equivalent to Conjecture 1, the original conjecture of Mills, Robbins, and Rumsey that has driven our story:

$$\frac{A_{n,r}}{A_{n,r+1}} = \frac{r(2n-r-1)}{(n-r)(n+r-1)}.$$

In Section 5.2 we proved that Conjecture 1, the refined alternating sign matrix conjecture, is equivalent to Conjecture 2:

$$A_{n,r} = \binom{n+r-2}{r-1} \frac{(2n-r-1)!}{(n-r)!} \prod_{j=0}^{n-2} \frac{(3j+1)!}{(n+j)!}.$$

The formula for the total number of $n \times n$ alternating sign matrices, Conjecture 3, then follows as a special case:

$$A_n = A_{n+1,1} = \prod_{j=0}^{n-1} \frac{(3j+1)!}{(n+j)!}.$$

To accomplish our task, we must evaluate the determinant in equation (7.17). Doron Zeilberger (1996b) came up with an ingenious method for accomplishing this. His method shows yet another approach to determinant evaluation, and it makes explicit connections to orthogonal polynomials, another thread that has been running under the surface of much that we have done.

Given a linear operator, T, defined on a ring of polynomials, T is uniquely determined by its **moments**, constants c_k defined by

$$c_k = T(x^k), \qquad k = 0, 1, 2, \dots .$$

If the matrix

$$\begin{pmatrix} c_0 & c_1 & \cdots & c_{n-1} \\ c_1 & c_2 & \cdots & c_n \\ \vdots & \vdots & & \vdots \\ c_{n-1} & c_n & \cdots & c_{2n-2} \end{pmatrix}$$

is non-singular for all values of n, then, as we shall see, there is a unique family of monic polynomials orthogonal with respect to T:

$$T\left(P_m(x)\, P_n(x)\right) = 0, \qquad m \neq n. \tag{7.23}$$

Let

$$P_n(x) = a_{n,0} + a_{n,1}x + \cdots + a_{n,n-1}x^{n-1} + x^n$$

where the $a_{n,j}$ are to be determined so that these polynomials satisfy equation (7.23). I leave it for you (exercise 7.3.1) to prove that this orthogonality condition is equivalent to

$$T\left(x^k\, P_n(x)\right) = 0, \qquad \text{for } 0 \le k \le n-1. \tag{7.24}$$

For each k in this interval, we must have

$$c_k\, a_{n,0} + c_{k+1}\, a_{n,1} + \cdots + c_{k+n-1}\, a_{n,n-1} + c_{k+n} = 0.$$

Orthogonality is equivalent to satisfying the following system of equations that uniquely determines the coefficients:

$$\begin{pmatrix} c_0 & c_1 & \cdots & c_{n-1} \\ c_1 & c_2 & \cdots & c_n \\ \vdots & \vdots & & \vdots \\ c_{n-1} & c_n & \cdots & c_{2n-2} \end{pmatrix} \begin{pmatrix} a_{n,0} \\ a_{n,1} \\ \vdots \\ a_{n,n-1} \end{pmatrix} = - \begin{pmatrix} c_n \\ c_{n+1} \\ \vdots \\ c_{2n-1} \end{pmatrix}.$$

Each polynomial has a unique representation as a linear combination of the $P_n(x)$. If $f(x)$ is a polynomial of degree d, then

$$f(x) = \sum_{k=0}^{d} \frac{T\left(f(x)\, P_k(x)\right)}{T\left(P_k(x)^2\right)}\, P_k(x).$$

If we know the moments for a given linear operator, then there is an explicit formula for the corresponding orthogonal polynomials as ratios of determinants:

$$Q_n(x) = (-1)^n \begin{vmatrix} 1 & x & \dots & x^n \\ c_0 & c_1 & \dots & c_n \\ \vdots & \vdots & & \vdots \\ c_{n-1} & c_n & \dots & c_{2n-1} \end{vmatrix} \Bigg/ \begin{vmatrix} c_0 & c_1 & \dots & c_{n-1} \\ c_1 & c_2 & \dots & c_n \\ \vdots & \vdots & & \vdots \\ c_{n-1} & c_n & \dots & c_{2n-2} \end{vmatrix}.$$

To see why this is so, we first observe that Q_n is a monic polynomial of degree n. Second, if we multiply $Q_n(x)$ by x^k and then apply the operator T, we get that

$$T\left(x^k Q_n(x)\right) = (-1)^n \frac{\begin{vmatrix} T(x^k) & T(x^{k+1}) & \dots & T(x^{n+k}) \\ c_0 & c_1 & \dots & c_n \\ \vdots & \vdots & & \vdots \\ c_{n-1} & c_n & \dots & c_{2n-1} \end{vmatrix}}{\begin{vmatrix} c_0 & c_1 & \dots & c_{n-1} \\ c_1 & c_2 & \dots & c_n \\ \vdots & \vdots & & \vdots \\ c_{n-1} & c_n & \dots & c_{2n-2} \end{vmatrix}},$$

which is zero for $0 \leq k \leq n-1$. Since the monic polynomials orthogonal with respect to T are unique, they must be equal to the Q_n.

We consider the family of orthogonal polynomials, $P_n(x)$, defined by the operator

$$T(x^k) = \frac{1 - q^{k+1}}{1 - q^{3(k+1)}}.$$

To evaluate our determinant, we apply the operator defined by

$$S(x^k) = \frac{1 - t^2 q^k}{1 - t^6 q^{3k}}$$

to the polynomial $P_{n-1}(x)$. When we combine the discussion we have just given with the definition of $N(n;t)$, we see that

$$\frac{\det N(n;t)}{\det N^*(n-1)} = (-1)^{n-1}\, S\left(P_{n-1}(x)\right), \tag{7.25}$$

where

$$N^*(n-1) \;=\; \left(\frac{1-q^{i+j-1}}{1-q^{3(i+j-1)}}\right)_{i,j=1}^{n-1}.$$

Restating the refined alternating sign matrix conjecture

We begin the next stage of the proof by focusing on the determinant of the matrix of moments associated with T, the matrix given by $N^*(n-1)$. We take equation (7.17) on page 242 and set $t^2 = \omega q^{1-n}$ so that on the left all summands except the last disappear. On the right, we expand the determinant as a sum over permutations. The product in front of the determinant has a factor of $1 - t^6 q^{3n-3}$ which is zero when $t^2 = \omega q^{1-n}$. The only terms that survive are those that take the factor of $(1-t^2q^{n-1})/(1-t^6q^{3n-3})$ from the determinant. In other words, we can replace the determinant by

$$(-1)^{n-1}\frac{1-t^2q^{n-1}}{1-t^6q^{3n-3}}\det N^*(n-1).$$

With $t^2 = \omega q^{1-n}$, equation (7.17) becomes

$$q^{-n(n-1)/4}(-\omega q)^{n-1}(1-\omega^2q^{1-n})(1-\omega^2q^{2-n})\cdots(1-\omega^2q^{-1})\,A_{n,n}(q)$$

$$= \frac{q^{-n(n-1)(5n-4)/6}}{(-3)^{(n-1)^2/2}}\prod_{j=1}^{n-1}\frac{1-q^{3j-3n}}{(1-\omega q^{j-n})(1-\omega q^{1-n-j})(1-q^{j-n})}$$

$$\times\frac{\prod_{i=2}^{n}\prod_{j=1}^{n}(1-q^{3i+3j-6})/(1-q^{i+j-2})}{\prod_{1\le i<j\le n-1}(1-q^{j-i})^2}(-1)^{n-1}\det N^*(n-1).$$

If we divide both sides by $(-1)^{n-1}\prod_{j=1}^{n-1}(1-\omega^2q^{j-n})$, this equality simplifies to

$$q^{-n(n-1)/4}(\omega q)^{n-1}\,A_{n,n}(q)$$

$$= \frac{q^{-n(n-1)(5n-4)/6}}{(-3)^{(n-1)^2/2}}\prod_{j=1}^{n-1}\frac{1}{1-\omega q^{1-n-j}}$$

$$\times\frac{\prod_{i=2}^{n}\prod_{j=1}^{n}(1-q^{3i+3j-6})/(1-q^{i+j-2})}{\prod_{1\le i<j\le n-1}(1-q^{j-i})^2}\det N^*(n-1). \tag{7.26}$$

At this point we pause to make the important observation that since $\lim_{q\to 1} A_{n,n}(q) = A_{n,n} \neq 0$, $A_{n,n}(q)$ cannot be zero, and so $N^*(n-1)$ is non-singular.

We now divide the left and right sides of equation (7.17) by the respective left and right sides of equation (7.26) and replace $N(n;t)/N^*(n-1)$ by $(-1)^{n-1} S\left(P_{n-1}(x)\right)$:

$$\begin{aligned}
\omega^n q^{1-n} \sum_{r=1}^{n} &(-1)^{r-1} q^{r-1} \omega^{2r} (1-\omega t^2)(1-\omega t^2 q)\cdots(1-\omega t^2 q^{r-2}) \\
&\times (1-\omega^2 t^2 q^r)\cdots(1-\omega^2 t^2 q^{n-1}) \frac{A_{n,r}(q)}{A_{n,n}(q)} \\
&= \prod_{j=1}^{n-1}(1-\omega q^{1-n-j}) \frac{\prod_{j=1}^{n}(1-t^6 q^{3j-3})/(1-t^2 q^{j-1})}{\prod_{j=2}^{n}(1-t^2 q^{1-j})(1-q^{1-j})} \\
&\qquad \times (-1)^{n-1} S\left(P_{n-1}(x)\right).
\end{aligned} \tag{7.27}$$

We have reduced the proof of the refined alternating sign matrix conjecture to finding the limit as q approaches 1 of the right-hand side of this equality, and then verifying that it is equal to the left-hand side with $q = 1$ and $A_{n,r}/A_{n,n}$ replaced by the conjectured value. It is time to investigate our particular linear operators in greater depth.

q-calculus

One of the simplest examples of a family of orthogonal polynomials was described by Adrien-Marie Legendre in 1784. For the linear transformation, we use the definite integral from -1 to 1:

$$c_k = \int_{-1}^{1} x^k \, dx = \begin{cases} 2/(k+1), & k \text{ even}, \\ 0, & k \text{ odd}. \end{cases}$$

The resulting **Legendre polynomials** are given by

$$\begin{aligned}
P_0(x) &= 1, \\
P_1(x) &= x, \\
P_2(x) &= x^2 - \frac{1}{3}, \\
P_3(x) &= x^3 - \frac{3}{5}x, \\
&\vdots
\end{aligned}$$

The linear operators that we need to use may look peculiar, but they are just discrete versions of the definite integral used to define the Legendre polynomials.

Before discussing this discrete integral, it is easiest to begin with the discrete derivative:

$$D_q f(x) = \frac{f(x) - f(xq)}{x - xq} = \frac{f(xq) - f(x)}{x(1-q)}. \tag{7.28}$$

The ordinary derivative is simply the limit as q approaches 1 of D_q. Fermat proved the rule for the derivative of an arbitrary rational power of x using this discrete version. We note that if a and b are positive integers, then

$$\frac{1-q^{a/b}}{1-q} = \frac{1-q^{a/b}}{1-q^{1/b}} \cdot \frac{1-q^{1/b}}{1-q^{b/b}} = \frac{1+q^{1/b}+\cdots+q^{(a-1)/b}}{1+q^{1/b}+\cdots+q^{(b-1)/b}},$$

and therefore

$$\begin{aligned} D\,x^{a/b} &= \lim_{q\to 1} \frac{x^{a/b} - (xq)^{a/b}}{x(1-q)} \\ &= x^{(a-b)/b} \lim_{q\to 1} \frac{1-q^{a/b}}{1-q} \\ &= \frac{a}{b}\, x^{(a-b)/b}. \end{aligned}$$

If a is negative, then we can replace $1 - q^{a/b}$ by $-q^{a/b}(1 - q^{-a/b})$, and the derivation still works.

The corresponding discrete integral is a summation over values in geometric progression where it now becomes necessary to insist that $|q| < 1$:

$$\begin{aligned} \int_0^a f(x)\, d_q x &= \sum_{j=0}^{\infty} f(aq^j)(aq^j - aq^{j+1}) \\ &= a(1-q) \sum_{j=0}^{\infty} f(aq^j) q^j, \end{aligned} \tag{7.29}$$

$$\begin{aligned} \int_a^b f(x)\, d_q x &= \int_0^b f(x)\, d_q x - \int_0^a f(x)\, d_q x \\ &= (1-q) \sum_{j=0}^{\infty} \left(b f(bq^j) - a f(aq^j) \right) q^j. \end{aligned} \tag{7.30}$$

It is left as an exercise to verify that this becomes the usual continuous integral in the limit as q approaches 1 from below, and that it satisfies its own fundamental theorem:

$$\begin{aligned} D_q \int_0^x f(t)\, d_q t &= f(x), \\ \int_a^b D_q\, f(x)\, d_q x &= f(b) - f(a). \end{aligned}$$

Discrete differentiation has a slightly modified product rule,

$$D_q\, [f(x)g(x)] = f(x)\, D_q\, g(x) + g(xq)\, D_q f(x).$$

It follows that we have a rule for integration by parts,

$$\int_a^b f(x)\, D_q\, g(x)\, d_q x = f(x)g(x)\bigg|_a^b - \int_a^b g(xq)\, D_q f(x)\, d_q x.$$

We now consider the linear operator defined by

$$T\,(f(x)) = \frac{1}{1-q^3} \int_q^1 f(x)\, d_{q^3} x.$$

As desired, we have that

$$\begin{aligned} T(x^k) &= \sum_{j=0}^{\infty} \left(q^{3jk} - q \cdot q^{(3j+1)k} \right) q^{3j} \\ &= \frac{1-q^{k+1}}{1-q^{3(k+1)}}. \end{aligned}$$

If we define α by $q^\alpha = t^2/q$, then the operator S is simply discrete integration against the weight function x^α:

$$\begin{aligned} S(x^k) &= \frac{1}{1-q^3} \int_q^1 x^k\, x^\alpha\, d_{q^3} x \\ &= \sum_{j=0}^{\infty} \left(q^{3j(k+\alpha)} - q \cdot q^{(3j+1)(k+\alpha)} \right) q^{3j} \\ &= \frac{1-q^{k+\alpha+1}}{1-q^{3(k+\alpha+1)}} \\ &= \frac{1-t^2 q^k}{1-t^6 q^{3k}}. \end{aligned}$$

Rodrigues

This work would be for naught unless we had an alternate description of the polynomials $P_n(x)$. In 1816, Olinde Rodrigues observed that the nth Legendre polynomial could be written as

$$\frac{n!}{(2n)!} D^n \left[(x^2-1)^n\right].$$

More generally, the polynomials

$$\frac{n!}{(2n)!} D^n \left[(x-a)^n(x-b)^n\right]$$

are monic and orthogonal with respect to the transformation $T(f(x)) = \int_a^b f(x)\,dx$. The orthogonality of these polynomials follows from the integration by parts formula and the fact that for $k < n$, the kth derivative of $[(x-a)^n(x-b)^n]$ is zero at both $x=a$ and $x=b$. For $m<n$, we have that

$$\begin{aligned}
&\int_a^b D^m\left[(x-a)^m(x-b)^m\right] D^n\left[(x-a)^n(x-b)^n\right] dx \\
&\quad = -\int_a^b D^{m+1}\left[(x-a)^m(x-b)^m\right] D^{n-1}\left[(x-a)^n(x-b)^n\right] dx \\
&\quad \vdots \\
&\quad = (-1)^m \frac{(2m)!}{m!} \int_a^b D^{n-m}\left[(x-a)^n(x-b)^n\right] dx \\
&\quad = (-1)^m \frac{(2m)!}{m!} D^{n-m-1}\left[(x-a)^n(x-b)^n\right]\Big|_a^b \\
&\quad = 0.
\end{aligned}$$

Benjamin Olinde Rodrigues was born in Bordeaux in 1794 to a prominent Sephardic family. He attended the École Normale and went on to the University of Paris from which he received his doctorate in 1816. His formula for Legendre polynomials was part of his doctoral thesis. He left mathematics to enter banking. History remembers him as an ardent supporter of the St. Simon movement, an early form of French socialism. He resumed his interest in mathematics in the late 1830s and made a significant contribution to the study of the group of isometries of Euclidean three-space (1840).

A formula analogous to the one discovered by Rodrigues works for the discrete version of the Legendre polynomials. If we define

$$P_n(x) = \frac{(1-q)^n}{(q^{n+1};q)_n} D_q^n \left[\prod_{i=0}^{n-1} (x - aq^i)(x - bq^i) \right],$$

then these are also monic polynomials that are orthogonal with respect to the discrete integral. Integration by parts works exactly as before except that x is replaced by xq in the function that gets integrated. The fact that for $m < n$,

$$D_q^{n-m-1} \left[\prod_{i=0}^{n-1} (xq^m - aq^i)(xq^m - bq^i) \right]$$

is zero at $x = a$ and at $x = b$ is left as an exercise. If $m < n$, then

$$\begin{aligned} \int_a^b D_q^m & \left[\prod_{i=0}^{m-1} (x - aq^i)(x - bq^i) \right] D_q^n \left[\prod_{i=0}^{n-1} (x - aq^i)(x - bq^i) \right] d_q x \\ &= (-1)^m \frac{(1-q)^m}{(q^{m+1};q)_m} D_q^{n-m-1} \left[\prod_{i=0}^{n-1} (xq^m - aq^i)(xq^m - bq^i) \right] \Bigg|_a^b \\ &= 0. \end{aligned}$$

We obtain the polynomials that we need for our determinant evaluation by replacing q by q^3 and then setting $a = q$ and $b = 1$.

Evaluation of $S(P_{n-1}(x))$

We have established that $S(P_{n-1}(x))$ is equal to the following discrete integral:

$$\frac{1}{1-q^3} \int_q^1 \frac{(1-q^3)^{n-1}}{(q^{3n};q^3)_{n-1}} D_{q^3}^{n-1} \left[\prod_{i=0}^{n-2} (x - q^{3i+1})(x - q^{3i}) \right] x^\alpha \, d_{q^3} x. \tag{7.31}$$

We perform integration by parts $n-1$ times and use the fact that

$$D_{q^3}^{n-1} x^\alpha = \frac{(q^{3(\alpha-n+2)};q^3)_{n-1}}{(1-q^3)^{n-1}} x^{\alpha-n+1}$$

to rewrite the integral:

$$\begin{aligned} S(P_{n-1}(x)) &= (-1)^{n-1} \frac{(q^{3(\alpha-n+2)};q^3)_{n-1}}{(1-q^3)(q^{3n};q^3)_{n-1}} \\ &\quad \times \int_q^1 \left[\prod_{i=0}^{n-2} (xq^{3n-3} - q^{3i+1})(xq^{3n-3} - q^{3i}) \right] x^{\alpha-n+1} \, d_{q^3} x. \end{aligned}$$

We now replace the integral by the sum that defines it, equation (7.30),

$$\begin{aligned} S(P_{n-1}(x)) &= (-1)^{n-1}\frac{(q^{3(\alpha-n+2)};q^3)_{n-1}}{(q^{3n};q^3)_{n-1}} \\ &\times \sum_{j=0}^{\infty}\left\{\left[\prod_{i=0}^{n-2}(q^{3n+3j-3}-q^{3i+1})(q^{3n+3j-3}-q^{3i})\right]q^{(3j)(\alpha-n+1)+3j}\right. \\ &- \left.\left[\prod_{i=0}^{n-2}(q^{3n+3j-2}-q^{3i+1})(q^{3n+3j-2}-q^{3i})\right]q^{(3j+1)(\alpha-n+1)+3j+1}\right\}. \end{aligned}$$

We replace q^α by t^2/q and then rewrite these summations using the rising q-factorial notation:

$$\begin{aligned} S(P_{n-1}(x)) &= (-1)^{n-1}\frac{(t^6q^{3-3n};q^3)_{n-1}}{(q^{3n};q^3)_{n-1}}\,q^{(n-1)(3n-5)} \\ &\times\left\{\sum_{j=0}^{\infty}(q^{3j+2};q^3)_{n-1}(q^{3j+3};q^3)_{n-1}(t^6q^{3-3n})^j\right. \\ &- \left. t^2q^{1-n}\sum_{j=0}^{\infty}(q^{3j+4};q^3)_{n-1}(q^{3j+3}:q^3)_{n-1}\right\}. \end{aligned}$$

These sums can be written as basic hypergeometric series:

$$\begin{aligned} S(P_{n-1}(x)) &= (-1)^{n-1}\frac{(t^6q^{3-3n};q^3)_{n-1}}{(q^{3n};q^3)_{n-1}}\,q^{(n-1)(3n-5)} \\ &\times\left\{(q^2;q^3)_{n-1}(q^3;q^3)_{n-1}\,{}_2\phi_1\left[\begin{matrix} q^{3n-1},\ q^{3n} \\ q^2 \end{matrix};q^3;t^6q^{3-3n}\right]\right. \\ &- \left.t^2q^{1-n}(q^3;q^3)_{n-1}(q^4;q^3)_{n-1}\,{}_2\phi_1\left[\begin{matrix} q^{3n},\ q^{3n+1} \\ q^4 \end{matrix};q^3;t^6q^{3-3n}\right]\right\}. \end{aligned}$$

We now apply the transformation formula given in equation (5.16) on page 170 to each of our basic hypergeometric series:

$$\begin{aligned} S(P_{n-1}(x)) &= (-1)^{n-1}\frac{(t^6q^{3-3n};q^3)_{n-1}}{(q^{3n};q^3)_{n-1}}\,q^{(n-1)(3n-5)}\,\frac{(t^6q^{3n};q^3)_\infty}{(t^6q^{3-3n};q^3)_\infty} \\ &\times\left\{(q^2;q^3)_{n-1}(q^3;q^3)_{n-1}\,{}_2\phi_1\left[\begin{matrix} q^{3-3n},\ q^{2-3n} \\ q^2 \end{matrix};q^3;t^6q^{3n}\right]\right. \\ &- \left.t^2q^{1-n}(q^3;q^3)_{n-1}(q^4;q^3)_{n-1}\,{}_2\phi_1\left[\begin{matrix} q^{3-3n},\ q^{4-3n} \\ q^4 \end{matrix};q^3;t^6q^{3n}\right]\right\}. \end{aligned}$$

We now substitute this back into equation (7.27) and take the limit of each side as q approaches 1:

$$\omega^n \sum_{r=1}^{n} (-1)^{r-1} \omega^{2r} (1-\omega t^2)^{r-1} (1-\omega^2 t^2)^{n-r} \frac{A_{n,r}}{A_{n,n}}$$
$$= \frac{(-3)^{n-1}(1-\omega)^{n-1}}{(1-t^2)^{2n-1}(n)_{n-1}}$$
$$\times \left\{ (2/3)_{n-1}\, {}_2F_1 \left[\begin{matrix} 1-n,\ 2/3-n \\ 2/3 \end{matrix} ; t^6 \right] \right.$$
$$\left. - t^2 (4/3)_{n-1}\, {}_2F_1 \left[\begin{matrix} 1-n,\ 4/3-n \\ 4/3 \end{matrix} ; t^6 \right] \right\}. \qquad (7.32)$$

We replace the index of summation, r, by $r+1$ and observe that the conjectured formula for $A_{n,r+1}/A_{n,n}$ is equal to

$$\frac{(1-n)_r (n)_r}{r!(2-2n)_r}.$$

We make this substitution into equation (7.32) and rewrite the summation on r as a hypergeometric series. We then make the observation that $1-\omega = -\omega(1-\omega^2)$. The refined alternating sign matrix conjecture has been reduced to verification of the following transformation formula:

$$(1-\omega^2 t^2)^{n-1}\, {}_2F_1 \left[\begin{matrix} 1-n,\ n \\ 2-2n \end{matrix} ; -\omega^2 \frac{1-\omega t^2}{1-\omega^2 t^2} \right]$$
$$= \frac{3^{n-1}(1-\omega^2)^{n-1}}{(1-t^2)^{2n-1}(n)_{n-1}}$$
$$\times \left\{ (2/3)_{n-1}\, {}_2F_1 \left[\begin{matrix} 1-n,\ 2/3-n \\ 2/3 \end{matrix} ; t^6 \right] \right.$$
$$\left. - t^2 (4/3)_{n-1}\, {}_2F_1 \left[\begin{matrix} 1-n,\ 4/3-n \\ 4/3 \end{matrix} ; t^6 \right] \right\}. \qquad (7.33)$$

We have emerged into well-traveled territory. Equation (7.33) is a cubic transformation formula for hypergeometric series. Two options are available. One is to check the literature to see whether this identity is known. It is. It is equation (38) on page 113 of Chapter 2, Section 11 of the first volume of *Higher Transcendental Functions*,† better known as the Bateman Project (Bateman 1953–55). Another approach is to turn to one of the computer packages available for proving identities involving hypergeometric series. These include Doron Zeilberger's `Ekhad` for

† This was pointed out to me by Dennis Stanton.

Maple and Peter Paule and Markus Schorn's `Zb` written for *Mathematica*. What the computer does is to find a recursive relation satisfied by the hypergeometric series. In our case, each side of equation (7.33) satisfies the same recursion. In fact, each of the three summations satisfies this recursive formula. If $f(n)$ is the respective function of n, then

$$\begin{aligned} 0 &= 3(9n^2-1)(1+t^2+t^4)^2 f(n) \\ &\quad - 6(1-\omega)(4n^2-1)(1+t^6) f(n+1) \\ &\quad - 4\omega(4n^2-1)(1-t^2)^2 f(n+2). \end{aligned} \tag{7.34}$$

Equation (7.33) now follows from the fact that it is correct for $n=1$ and for $n=2$.

Q.E.D.

Exercises

7.3.1 Prove that $\{P_n(x)\}_{n=0}^{\infty}$ is orthogonal with respect to the linear operator T if and only if $T\left(x^k P_n(x)\right) = 0$ for $0 \le k < n$.

7.3.2 Verify that

$$\lim_{q\to 1^-} \int_0^a f(x)\, d_q x = \int_0^a f(x)\, dx.$$

7.3.3 Verify that

$$D_q \int_0^x f(t)\, d_q t = f(x)$$

and that

$$\int_a^b D_q f(x)\, d_q x = f(b) - f(a).$$

7.3.4 Verify the product rule for q-differentiation.

7.3.5 Verify that for $m < n$,

$$D_q^{n-m-1}\left[\prod_{i=0}^{n-1}(xq^m - aq^i)(xq^m - bq^i)\right]$$

is zero at $x = a$ and at $x = b$.

7.3.6 Verify that equation (7.34) is correct for $n=1$ and $n=2$.

7.4 Forward

Our story has reached its end. We have proven that the ratios of adjacent values of $A_{n,r}$ can be found by adding numerators and adding denominators of the two ratios diagonally above.

Of course, this is not really an end at all, but a beginning. New connections and new tools have been discovered. Methods for evaluating determinants with polynomial entries have already gone well beyond what we have seen in this book. The Yang–Baxter equation has become the object of serious study by combinatorialists, many of whom are now adept at applying it to a variety of problems. I have totally omitted reference to the field of tiling problems that have come out of the Mills, Robbins, and Rumsey conjectures, initially as an alternate way of stating them, but now as an endeavor that has taken on a life of its own with its own set of conjectures and techniques and results. I have barely touched on the ties to the theory of orthogonal polynomials and the q-calculus, of connections with elliptic modular functions, of the role these ideas play in the continuing development of representation theory and Lie algebras.

These fields are too new for there to be much literature outside of journal articles available for those who wish to take the next step, but the books that are likely to be most helpful are *Enumerative Combinatorics* (Stanley 1986c, 1999) and the second edition of *Symmetric Functions and Hall Polynomials* (Macdonald 1995).

Collapse of a metaphor

Having reached the end of our story, I want to draw some conclusions about the nature and purpose of proof. I began with and throughout this book continued to make allusion to the metaphor of climbing mountains and exploring new territory. It is a useful analogy that is popular among mathematicians because it conveys a spirit of discovery and adventure. But there are problems when one attempts to push this metaphor too far.

Mathematical results are not isolated peaks which, once conquered, lose their interest. Moreover, the adventure does not lie in setting out from known territory and finding one's way to the distant peak. Rather, as we saw in the proof of the refined alternating sign matrix conjecture, we start in unfamiliar territory and end when we have reduced our problem to one that is recognizable, that others have solved. The doing of mathematics is more akin to being dropped on a distant and unknown mountain peak and then seeking to find one's way home. This image contains an important truth. Research in mathematics almost never begins with careful definitions and lemmas on which we build until something interesting is discovered. It starts with the discovery, and proof is the process of tying that discovery back to what is already known.

There are still difficulties with the metaphor of exploration in whichever direction it may be taken. It does not adequately convey the importance of connections or the value of a new and truly different proof of a long-established result. It does not communicate the fact that each new proof is more than an individual accomplishment or a personal opportunity to hone skills. A proof is also a community asset that illuminates other problems. Furthermore, our metaphor suggests that mathematics is purely a science of discovery when in fact it is also an act of creation.

I would like to conclude by proposing a different metaphor, one with its own limitations and imperfections but which may help to complement the images presented so far. I would like to consider the doing of mathematics and the finding of proofs as analogous to the work of the archæologist. When Mills, Robbins, and Rumsey first discovered their conjecture, they were not dissimilar to the archæologist who has just unearthed a strange and marvelous object of unknown provenance and purpose. What is it? What was it used for? Why is it here? What does it tell us about the people who once lived here? The real work of the archæologist is to make connections: connections to other objects at other places at other times, connections to other facts that are known about this particular site. The goal of the archæologist is to provide a context in which we can understand this object. As each object comes to be understood, it facilitates the interpretation of others, not just in this place, but also in other places and from other times. It provides a foundation upon which we construct our theories.

This is the role of proof, to enrich the entire web of context that leads to understanding. The mathematician does not dig for lost artifacts of a vanished civilization but for the fundamental patterns that undergird our universe, and like the archæologist we usually find only small fragments. As archæology attempts to reconstruct the society in which this object was used, so mathematics is the reconstruction of these patterns into terms that we can comprehend.

This book has been about proofs and confirmations, the process of taking a newly unearthed discovery and seeking the proof that will relate it to what is known. It is one aspect of the process that creates mathematics. The explanation of the use and significance of an object does not end its archæological interest. This knowledge now allows it to move into the greater debate of how we are to understand particular communities or civilizations. So too in mathematics, the proof of the refined alternating sign matrix conjecture is the ground from which we can begin to seek its true significance. At the next stage, we seek the theories that can explain what we have seen and predict the directions that should be most fruitful.

Bibliography

Andrews, George E. 1974. Applications of basic hypergeometric functions. *SIAM Review* **16**: 441–484.

———. 1975. Problems and prospects for basic hypergeometric functions. In *Theory and Application of Special Functions*, ed. R. Askey. Pp. 191–224. New York: Academic Press.

———. 1976. *The Theory of Partitions*. Reading, Mass.: Addison-Wesley. Reprint. Cambridge: Cambridge University Press, 1998.

———. 1977. Plane partitions (II): The equivalence of the Bender-Knuth and MacMahon conjecture. *Pacific Journal of Mathematics* **72**: 283–291.

———. 1978. Plane partitions (I): The MacMahon conjecture. *Studies in Foundations and Combinatorics, Advances in Mathematics Supplementary Studies* **1**: 131–150.

———. 1979. Plane partitions (III): The weak Macdonald conjecture. *Inventiones Mathematicae* **53**: 193–225.

———. 1980. Notes on the Dyson conjecture. *SIAM Journal on Mathematical Analysis* **11**: 787–792.

———. 1987. Plane partitions. IV. A conjecture of Mills-Robbins-Rumsey. *Aequationes Mathematicae* **33**: 230–250.

———. 1994. Plane partitions. V. The TSSCPP conjecture. *Journal of Combinatorial Theory, Series A* **66**: 28–39.

Andrews, George E., Richard Askey, and Ranjan Roy. 1999. *Special Functions*. Cambridge: Cambridge University Press.

Askey, Richard A., ed. 1975. *Orthogonal Polynomials and Special Functions*, Regional Conference Series in Applied Mathematics **21**. Philadelphia: Society for Industrial and Applied Mathematics.

———. 1980. Some basic hypergeometric extensions of integrals of Selberg and Andrews. *SIAM Journal on Mathematical Analysis* **11**: 938–951.

———. 1988. How can mathematicians and mathematical historians help each other? In *History and Philosophy of Modern Mathematics*, ed. William Aspray and Philip Kitcher. *Minnesota Studies in the Philosophy of Science* **XI**. Pp. 201–220. Minneapolis: University of Minnesota Press.

Bateman Manuscript Project. 1953–55. *Higher Transcendental Functions*. 3 vols. New York: McGraw-Hill.

Baxter, Rodney J. 1971. Eight-vertex model in lattice statistics. *Physical Review Letters* **26**: 832–833.

Baxter, Rodney J. 1980. Hard hexagons: Exact solution. *Journal of Physics A: Mathematics and General* **13**: L61–L70.

———. 1981. Rogers-Ramanujan identities in the hard hexagon model. *Journal of Statistical Physics* **26**: 427–452.

———. 1982. *Exactly Solved Models in Statistical Mechanics*. London: Academic Press.

Bender, E. A., and D. Knuth. 1972. Enumeration of plane partitions. *Journal of Combinatorial Theory* **13**: 40–54.

Bressoud, David M. 1983. An easy proof of the Rogers-Ramanujan identities. *Journal of Number Theory* **16**: 235–241.

———. 1994. *A Radical Approach to Real Analysis*. Washington, DC: The Mathematical Association of America.

———. 1998. Elementary proof of MacMahon's conjecture. *Journal of Algebraic Combinatorics* **7**: 253–257.

Carlitz, L. 1967. Rectangular arrays and plane partitions. *Acta Arithmetica* **13**: 29–47.

Carlitz, L., and M. V. Subbarao. 1972. On a combinatorial identity of Winquist and its generalization. *Duke Mathematical Journal* **39**: 165–172.

Carter, Roger W. 1989. *Simple Groups of Lie Type.* London: John Wiley & Sons.

Cauchy, Augustin-Louis. 1815. Mémoire sur les fonctions qui ne peuvent obtenir que deux valeurs égales et de signes contraires par suite des transpositions opérées entre les variables qu'elles renferment. *Journal de l'École Polytechnique* **10**, Cahier 17: 29–112. Reprinted in *Œuvres complètes d'Augustin Cauchy* series 2, Vol. 1, pp. 91–161. Paris: Gauthier-Villars, 1899.

———. 1829. Sur l'équation à l'aide de laquelle on détermine les inégalités séculaires des mouvements des planètes. *Exercises de Mathématiques* **4**: 140–160. Reprinted in *Œuvres complètes d'Augustin Cauchy* series 2, Vol. 9, 174–195. Paris: Gauthier-Villars, 1899.

Cayley, Arthur. 1848. Sur les déterminants gauches. *Journal für die reine und angewandte Mathematik* **38**: 93–96. Reprinted in *The Collected Mathematical Papers of Arthur Cayley.* Vol. 1, pp. 410–413, 1889.

———. 1883. Presidential address to the British Association, *Report of the British Association for the Advancement of Science.* Pp. 3–37. Reprinted in *The Collected Mathematical Papers of Arthur Cayley.* Vol. 11, pp. 429–459, 1896.

Desnanot, P. 1819. *Complément de la théorie des équations du premier degré*. Paris.

Dodgson, Charles L. 1866. Condensation of determinants. *Proceedings of the Royal Society, London* **15**: 150–155.

Doran, William F., IV. 1993. A connection between alternating sign matrices and totally symmetric self-complementary plane partitions. *Journal of Combinatorial Theory, Series A* **64**: 289–310.

Duffin, R. J. 1956. Basic properties of discrete analytic functions. *Duke Mathematical Journal*. **23**: 335–363.

Dyson, Freeman J. 1962. Statistical theory of the energy levels of complex systems I. *Journal of Mathematical Physics* **3**: 140–156.

———. 1972. Missed opportunities. *Bulletin of the American Mathematical Society* **78**: 635–652.

Fasenmyer, Sister Mary Celine. 1945. Some generalized hypergeometric polynomials. Ph.D. dissertation. University of Michigan.

Fulton, William. 1997. *Young Tableaux*. London Mathematical Society Student Texts **35**. Cambridge: Cambridge University Press.

Gasper, George, and Mizan Rahman. 1990. *Basic Hypergeometric Series*. Cambridge: Cambridge University Press.

Gessel, Ira. 1979. Tournaments and Vandermonde's determinant. *Journal of Graph Theory* **3**: 305–307.

Gessel, Ira, and Dennis Stanton. 1982. Strange evaluations of hypergeometric series. *SIAM Journal on Mathematical Analysis* **13**: 295–308.

Gessel, Ira, and Gérard Viennot. 1985. Binomial determinants, paths, and hook length formulae. *Advances in Mathematics* **58**, no. 3: 300–321.

Good, I. J. 1970. Short proof of a conjecture of Dyson. *Journal of Mathematical Physics* **11**: 1884.

Gordon, Basil. 1971a. Notes on plane partitions. IV. Multirowed partitions and strict decrease along columns. *Proceedings of Symposium in Pure Mathematics* **19**: 91–100. Providence, RI: American Mathematical Society.

———. 1971b. Notes on plane partitions. V. *J. Combinatorial Theory Series B* **11**: 157–168.

Gordon, B., and L. Houten. 1968a. Notes on plane partitions. I. *Journal of Combinatorial Theory* **4**: 72–80.

———. 1968b. Notes on plane partitions. II. *Journal of Combinatorial Theory* **4**: 81–99.

———. 1969. Notes on plane partitions. III. *Duke Mathematical Journal* **36**: 801–824.

Gosper, R. William, Jr. 1978. Decision procedure for indefinite hypergeometric summation. *Proceedings of the National Academy of Science USA* **75**: 40–42.

Heine, Heinrich Eduard. 1846. Über die Reihe $1+\frac{(q^\alpha-1)(q^\beta-1)}{(q-1)(q^\gamma-1)}\cdot x+\frac{(q^\alpha-1)(q^{\alpha+1}-1)(q^\beta-1)(q^{\beta+1}-1)}{(q-1)(q^2-1)(q^\gamma-1)(q^{\gamma+1}-1)}\cdot x^2+\cdots$. *Journal fur die Reine und Angewandt Mathematik* **32**: 210–212.

———. 1847. Untersuchungen über die Reihe $1+\frac{(q^\alpha-1)(q^\beta-1)}{(q-1)(q^\gamma-1)}\cdot x+\frac{(q^\alpha-1)(q^{\alpha+1}-1)(q^\beta-1)(q^{\beta+1}-1)}{(q-1)(q^2-1)(q^\gamma-1)(q^{\gamma+1}-1)}\cdot x^2+\cdots$. *Journal fur die Reine und Angewandt Mathematik* **34**: 285–328.

Horgan, John. 1993. The death of proof. *Scientific American* **269**: 92–103.

Horn, J. 1889. Ueber die Convergenz der hypergeometrischen Reihen zweier und dreier Veränderlichen. *Mathematische Annalen* **34**: 544–600.

Izergin, Anatoli G. 1987. Partition function of a six-vertex model in a finite volume. (Russian) *Dokl. Akad. Nauk SSSR* **297**: 331–333.

Jacobi, C. G. J. 1827. Ueber die Pfaffsche Methode. *Journal fur die Reine und Angewandt Mathematik* **2**: 347–357. Reprinted in *C. G. J. Jacobi: Gesammelte Werke*. Vol. 4, pp. 17–29. Berlin: Georg Reimer, 1884.

———. 1833. De binis quibuslibet functionibus homogeneis secundi ordinis per substitutiones lineares in alias binas transformandis. *Journal fur die Reine und Angewandt Mathematik*. **12**: 1–69. Reprinted in *C. G. J. Jacobi: Gesammelte Werke*. Vol. 3, pp. 191–268. Berlin: Georg Reimer, 1884.

———. 1841a. De formatione et proprietatibus Determinantium. *Journal fur die Reine und Angewandt Mathematik*. **22**: 285–318. Reprinted in *C. G.*

J. Jacobi: Gesammelte Werke. Vol. 3, pp. 355–392. Berlin: Georg Reimer, 1884.

Jacobi, C. G. J. 1841b. De determinantibus functionalibus. *Journal fur die Reine und Angewandt Mathematik* **22**: 319–359. Reprinted in *C. G. J. Jacobi: Gesammelte Werke*. Vol. 3, pp. 393–438. Berlin: Georg Reimer, 1884.

———. 1841c. De functionibus alternantibus earunque divisione per productum e differentiis elementorum conflatum. *Journal fur die Reine und Angewandt Mathematik* **22**: 360–371. Reprinted in *C. G. J. Jacobi: Gesammelte Werke*. Vol. 3, pp. 439–452. Berlin: Georg Reimer, 1884.

———. 1844-1845. Theoria novi multiplicatoris systemati æquationum differentialium vulgarium applicandi. *Journal fur die Reine und Angewandt Mathematik* **27**: 199-268; **29**: 213–279, 333–376. Reprinted in *C. G. J. Jacobi: Gesammelte Werke*. Vol. 4, pp. 317–509. Berlin: Georg Reimer, 1884.

Kac, M., G.-C. Rota, and J. T. Schwartz. 1992. *Discrete Thoughts: Essays on Mathematics, Science, and Philosophy*. 2nd edition. Boston: Birkhauser.

Knuth, Donald E. 1968. *The Art of Computer Programming*. Vol. 1: *Fundamental Algorithms*. Reading, Mass.: Addison-Wesley.

———. 1969. *The Art of Computer Programming*. Vol. 2: *Seminumerical Algorithms*. Reading, Mass.: Addison-Wesley.

———. 1973. *The Art of Computer Programming*. Vol. 3: *Sorting and Searching*. Reading, Mass.: Addison-Wesley.

Korepin, V. E., N. M. Bogoliubov, and A. G. Izergin. 1993. *Quantum Inverse Scattering Method and Correlation Functions*. Cambridge: Cambridge University Press.

Krattenthaler, Christian. 1990. Generating functions for plane partitions of a given shape. *Manuscripta Mathematica* **69**: 173–201.

———. 1997. Determinant identities and a generalization of the number of totally symmetric self-complementary plane partitions. *Electronic Journal of Combinatorics* **4**, no. 1, paper #27, 62 pp.

Kuperberg, Greg. 1994. Symmetries of plane partitions and the permanent-determinant method. *Journal of Combinatorial Theory, Series A* **68**: 115–151.

———. 1996a. Four symmetry classes of plane partitions under one roof. *Journal of Combinatorial Theory, Series A* **75**: 295–315.

———. 1996b. Another proof of the alternating sign matrix conjecture. *International Mathematics Research Notes* **1996**: 139–150.

Lagrange, Joseph-Louis. 1773. Recherches d'arithmétiques. *Nouvelle Mémoire de l'Académie Royale (de Berlin)* 265–312.

Lakatos, Imre. 1976. *Proofs and Refutations: The Logic of Mathematical Discovery*. Cambridge: Cambridge University Press.

Lindström, Bernt. 1973. On the vector representations of induced matroids. *Bulletin of the London Mathematical Society* **5**: 85–90.

Littlewood, D. E. 1950. *The Theory of Group Characters*. Oxford: Oxford University Press.

Macdonald, Ian G. 1972. Affine roots systems and Dedekind's η-function. *Inventiones Mathematicae*. **15**: 91–143.

———. 1979. *Symmetric Functions and Hall Polynomials*. Oxford: Oxford University Press.

———. 1995. *Symmetric Functions and Hall Polynomials*. 2d ed. Oxford:

Oxford University Press.

MacMahon, Percy A. 1897. Memoir on the theory of partitions of numbers – Part I. *Philosophical Transactions of the Royal Society of London* **187**: 619–673. Reprinted in *Percy Alexander MacMahon: Collected Papers*, ed. George E. Andrews. Vol. 1, pp. 1026–1080. Cambridge, Mass.: MIT Press, 1978.

———. 1898. James Joseph Sylvester. *Proceedings of the Royal Society* **63**: ix–xxv. Reprinted in *Percy Alexander MacMahon: Collected Papers*, ed. George E. Andrews. Vol. 2, pp. 1384–1405. Cambridge, Mass.: MIT Press, 1978.

———. 1899. Partitions of numbers whose graphs possess symmetry. *Transactions Cambridge Philosophical Society* **17**: 149–170. Reprinted in *Percy Alexander MacMahon: Collected Papers*, ed. George E. Andrews. Vol. 1, pp. 867–883. Cambridge, Mass.: MIT Press, 1978.

———. 1912a. Memoir on the theory of partitions of numbers – Part V. Partitions in two-dimensional space, to which is added an adumbration of the theory of partitions in three-dimensional space. *Philosophical Transactions of the Royal Society of London* **211**: 75–110. Reprinted in *Percy Alexander MacMahon: Collected Papers*, ed. George E. Andrews. Vol. 1, pp. 1328–1363. Cambridge, Mass.: MIT Press, 1978.

———. 1912b. Memoir on the theory of partitions of numbers – Part VI. Partitions in two-dimensional space. *Philosophical Transactions of the Royal Society of London* **211**: 345–373. Reprinted in *Percy Alexander MacMahon: Collected Papers*, ed. George E. Andrews. Vol. 1, pp. 1404–1434. Cambridge, Mass.: MIT Press, 1978.

———. 1915–1916. *Combinatory Analysis*. 2 vols. Cambridge University Press. Reprint. New York: Chelsea, 1960.

———. 1922. The connexion between the sum of the squares of the divisors and the number of partitions of a given number. *Messenger of Mathematics* **52**: 113–116. Reprinted in *Percy Alexander MacMahon: Collected Papers*, ed. George E. Andrews. Vol. 1, pp. 1364–1367. Cambridge, Mass.: MIT Press, 1978.

Mills, W. H., D. P. Robbins, and H. Rumsey. 1982. Proof of the Macdonald conjecture. *Inventiones Mathematicae* **66**: 73–87.

———. 1983. Alternating sign matrices and descending plane partitions. *Journal of Combinatorial Theory, Series A* **34**: 340–359.

———. 1986. Self-complementary totally symmetric plane partitions. *Journal of Combinatorial Theory, Series A* **42**: 277-292.

———. 1987. Enumeration of a symmetry class of plane partitions. *Discrete Mathematics* **67**: 43–55.

Minding, Ferdinand. 1829. Auflösung einiger Aufgaben der analytischen Geometrie vermittelst des barycentrischen Calculs. *Journal für die Reine und Angewandt Mathematik* **5**: 397–401.

Morris, William G., II. 1982. Constant term identities for finite and affine root systems: Conjectures and theorems. Ph.D. dissertation. University of Wisconsin, Madison.

Muir, Thomas. 1882. *A Treatise on the Theory of Determinants*. London: MacMillan and Co.

———. 1906–23. *The Theory of Determinants in the Historical Order of Development*. 4 vols. London: Macmillan and Co.

———. 1930. *Contributions to the History of Determinants, 1900-1920*.

London: Blackie & Son.

O'Hara, Kathleen M. 1990. Unimodality of Gaussian coefficients: A constructive proof. *Journal of Combinatorial Theory, Series A* **53**: 29–52.

Okada, Soichi. 1989. On the generating functions for certain classes of plane partitions. *Journal of Combinatorial Theory, Series A* **51**: 1–23.

———. 1993. Alternating sign matrices and some deformations of Weyl's denominator formulas. *Journal of Algebraic Combinatorics* **2**: 155–176.

Onsager, Lars. 1944. Chrystal Statistics I: A two-dimensional model with an order-disorder transition. *Physical Review* **65**: 117–149.

Parshall, Karen Hunger, and David E. Rowe. 1994. *The Emergence of the American Mathematical Research Community, 1876–1900: J. J. Sylvester, Felix Klein, and E. H. Moore*. Providence, RI: American Mathematical Society.

Petkovšek, Marko, Herbert S. Wilf, and Doron Zeilberger. 1996. $A = B$. Wellesley, Massachusetts: A K Peters.

Pfaff, Johann Friedrich. 1797. Observationes analyticæ ad L. Euler Institutiones Calculi Integralis, Vol. IV, Supplem. II et IV, Historia de 1793, *Nova acta acad. sci. Petropolitanæ* **11**: 38–57.

———. 1814–15. Methodus generalis, æquationes differentiarum partialium, nec non æquationes differentiales vulgares, utrasque primi ordinis, inter quotcunque variables, completi integrandi. *Abhandlung der Akademie der Wissenschaften (mathematische klasse) Berlin*. Pp. 76–136.

Popper, Karl Raimund. 1963. *Conjectures and Refutations: The Growth of Scientific Knowledge*. London: Routledge & K. Paul.

Proctor, Robert A. 1983. Shifted plane partitions of trapezoidal shape. *Proceedings of the American Mathematical Society* **89**: 553–559.

———. 1988. Odd sympletic groups. *Inventiones Mathematicae* **92**: 307–332.

Rademacher, Hans. 1964. *Lectures on Elementary Number Theory*. New York: Blaisdell. Reprint. Huntingdon, NY: Krieger, 1977.

Ramanujan, S. 1919. Proof of certain identities in combinatory analysis. *Proceedings Cambridge Philosophical Society* **19**: 214–216. Reprinted in *Collected Papers of Srinivasa Ramanujan*. Pp. 214–215. Cambridge University Press, 1927. Reprint. New York: Chelsea, 1962.

Rodrigues, B. Olinde. 1815. Mémoire sur l'attraction des sphéroides. *École Polytechn. Corresp.* **3**: 361–385.

Rogers, L. J. 1894. Second memoir on the expansion of certain infinite products. *Proceedings London Mathematical Society* **25**: 318–343.

Robbins, David P. 1991. The story of 1, 2, 7, 42, 429, 7436, ... *The Mathematical Intelligencer* **13**: 12–19.

Robbins, David P., and Howard Rumsey. 1986. Determinants and alternating sign matrices. *Advances in Mathematics* **62**: 169–184.

Schur, I. J. 1917. Ein Beitrag zur additiven Zahlentheorie und zur Theorie der Kennenbrüche. *S.-B. Preuss. Akad. Wiss. Phys.-Math. Kl.*. 302–321. Reprinted in *Gesammelte Abhandlungen*. Vol. 2, pp. 117-136. Berlin: Springer-Verlag, 1973.

———. 1918. Aufgabe # 569. *Archiv der Mathematik und Physik (3)* **27**: 163. Reprinted in *Gesammelte Abhandlungen*. Vol. 3, p. 456. Berlin: Springer-Verlag, 1973.

Stanley, Richard. 1971. Theory and application of plane partitions. I, II. *Studies in Applied Mathematics* **50**: 167–188, 259–279.

———. 1981. Review of *Symmetric Functions and Hall Polynomials. Bulletin of the American Mathematical Society* **4**: 254–265.

———. 1986a. A baker's dozen of conjectures concerning plane partitions. In *Combinatoire énumérative*, ed. G. Labelle and P. Leroux. *Lecture Notes in Mathematics* **1234**: 285–293. New York: Springer-Verlag.

———. 1986b. Symmetries of plane partitions. *Journal of Combinatorial Theory, Series A* **43**: 103–113.

———. 1986c. *Enumerative Combinatorics*. Vol. 1. Monterey, California: Wadsworth and Brooks/Cole. Reprint. Cambridge: Cambridge University Press, 1997.

———. 1999. *Enumerative Combinatorics*. Vol. 2. Cambridge: Cambridge University Press.

Stanton, Dennis. 1986. Sign variations of the Macdonald identities. *SIAM Journal on Mathematical Analysis* **17**: 1454–1460.

———. 1989. An elementary approach to the Macdonald identities. q-series and partitions. In *q-Series and Partitions*, ed. D. Stanton. *Institute for Mathematics and Its Applications* **18**: 139–149. New York: Springer-Verlag.

Stembridge, John. 1988. A short proof of Macdonald's conjecture for the root system of type A. *Proceedings of the American Mathematical Society* **102**: 777–786.

———. 1990a. Hall-Littlewood functions, plane partitions, and the Rogers-Ramanujan identities. *Transactions of the American Mathematical Society* **319**: 469–498.

———. 1990b. Nonintersecting paths, Pfaffians, and plane partitions *Advances in Mathematics* **83**: 96–131.

———. 1994. Some hidden relations involving the ten symmetry classes of plane partitions. *Journal of Combinatorial Theory, Series A* **68**: 372–409.

———. 1995. The enumeration of totally symmetric plane partitions. *Advances in Mathematics* **111**: 227–243.

Sylvester, James Joseph. 1882. A constructive theory of partitions, arranged in three acts, an interact, and an exodion. *American Journal of Mathematics* **5**: 251–330. Reprinted in *The Collected Mathematical Papers of J. J. Sylvester.* Vol. 4, pp. 1–81. Cambridge University Press, 1912. Reprint. New York: Chelsea, 1973.

———. 1883. Letter from Sylvester to Arthur Cayley, 16 March. Cited on page 129 of Parshall & Rowe, 1994.

———. 1878. Proof of the hitherto undemonstrated fundamental theorem of invariants. *Philosophical Magazine* **5**: 178–188. Reprinted in *The Collected Mathematical Papers of J. J. Sylvester.* Vol. 3, pp. 117-126. Cambridge University Press, 1912. Reprint. New York: Chelsea, 1973.

Trudi, Nicholas. 1864. Intorno ad un determinante piu generale di quello che suol dirsi determinante delle radici di una equazione, ed alle funzioni simmetriche complete di questi radici. *Rendic. dell'Accad.* (Napoli): 121–134.

Vandermonde, Alexandre Théophile. 1772. Mémoire sur des irrationnelles de différens ordres avec une application au cercle. *Mémoires de l'Académie Royale des Sciences, Paris*: 489–498.

Whittaker, E. T., and G. N. Watson. 1927. *A Course of Modern Analysis*. 4th ed. Cambridge: Cambridge University Press.

Wilf, Herbert S. 1994. *Generatingfunctionology*. 2nd ed. San Diego: Academic Press.

Wimp, Jet. 1997. Mathematics not. *The Mathematical Intelligencer* **19**: 70–75.

Winquist, Lasse. 1969. An elementary proof of $p(11m+6) \equiv 0 \pmod{11}$. *Journal of Combinatorial Theory* **6**: 56–59.

Zeilberger, Doron. 1980a. The algebra of linear partial difference operators and its applications. *SIAM Journal on Mathematical Analysis* **11**: 919–9324.

———. 1980b. Partial difference equations in $m_1 \geq \cdots \geq m_n \geq 0$ and their applications to combinatorics. *Discrete Mathematics* **31**: 65–77.

———. 1988. A unified approach to Macdonald's root-system conjectures. *SIAM Journal on Mathematical Analysis* **19**: 987–1013.

———. 1989. Kathy O'Hara's constructive proof of the unimodality of the Gaussian polynomials. *American Mathematical Monthly* **96**: 590–602.

———. 1990. A Stembridge-Stanton style proof of the Habsieger-Kadell q-Morris identity. *Discrete Mathematics* **79**: 313–322.

———. 1996a. Proof of the alternating sign matrix conjecture. *Electronic Journal of Combinatorics* **3** (1996), R13. Also published in *The Foata Festschrift*, ed. Jacques Désarménien, Adalbert Kerber, and Volker Strehl. Pp. 289–372. Gap, France: Imprimerie Louis-Jean.

———. 1996b. Proof of the refined alternating sign matrix conjecture. *New York Journal of Mathematics* **2**: 59–68.

Zeilberger, Doron, and David M. Bressoud. 1985. A proof of Andrews' q-Dyson conjecture. *Discrete Mathematics* **54**: 201–224.

Index of Notation

General Index